Lecture Notes in Mathematics

Edited by J.-M. Morel, F. Takens and B. Teissier

Editorial Policy
for the publication of monographs

1. Lecture Notes aim to report new developments in all areas of mathematics and their applications – quickly, informally and at a high level. Mathematical texts analysing new developments in modelling and numerical simulation are welcome.

   Monograph manuscripts should be reasonably self-contained and rounded off. Thus they may, and often will, present not only results of the author but also related work by other people. They may be based on specialised lecture courses. Furthermore, the manuscripts should provide sufficient motivation, examples and applications. This clearly distinguishes Lecture Notes from journal articles or technical reports which normally are very concise. Articles intended for a journal but too long to be accepted by most journals, usually do not have this „lecture notes" character. For similar reasons it is unusual for doctoral theses to be accepted for the Lecture Notes series, though habilitation theses may be appropriate.

2. Manuscripts should be submitted (preferably in duplicate) either to Springer's mathematics editorial in Heidelberg, or to one of the series editors (with a copy to Springer). In general, manuscripts will be sent out to 2 external referees for evaluation. If a decision cannot yet be reached on the basis of the first 2 reports, further referees may be contacted: The author will be informed of this. A final decision to publish can be made only on the basis of the complete manuscript, however a refereeing process leading to a preliminary decision can be based on a pre-final or incomplete manuscript. The strict minimum amount of material that will be considered should include a detailed outline describing the planned contents of each chapter, a bibliography and several sample chapters.

   Authors should be aware that incomplete or insufficiently close to final manuscripts almost always result in longer refereeing times and nevertheless unclear referees' recommendations, making further refereeing of a final draft necessary.

   Authors should also be aware that parallel submission of their manuscript to another publisher while under consideration for LNM will in general lead to immediate rejection.

3. Manuscripts should in general be submitted in English. Final manuscripts should contain at least 100 pages of mathematical text and should always include

   – a table of contents;

   – an informative introduction, with adequate motivation and perhaps some historical remarks: it should be accessible to a reader not intimately familiar with the topic treated;

   – a subject index: as a rule this is genuinely helpful for the reader.

   For evaluation purposes, manuscripts may be submitted in print or electronic form (print form is still preferred by most referees), in the latter case preferably as pdf- or zipped ps-files. Lecture Notes volumes are, as a rule, printed digitally from the authors' files. To ensure best results, authors are asked to use the LaTeX2e style files available from Springer's web-server at:

   ftp://ftp.springer.de/pub/tex/latex/mathegl/mono/ (for monographs) and

   ftp://ftp.springer.de/pub/tex/latex/mathegl/mult/ (for summer schools/tutorials).

   Additional technical instructions, if necessary, are available on request from lnm@springer-sbm.com.

Continued on inside back-cover

Lecture Notes in Mathematics

1924

Editors:
J.-M. Morel, Cachan
F. Takens, Groningen
B. Teissier, Paris

Michael Wilson

Weighted Littlewood-Paley Theory and Exponential-Square Integrability

 Springer

Author

Michael Wilson
Department of Mathematics
University of Vermont
Burlington, Vermont 05405
USA

e-mail: wilson@cems.uvm.edu

Library of Congress Control Number: 2007934022

Mathematics Subject Classification (2000): 42B25, 42B20, 42B15, 35J10

ISSN print edition: 0075-8434
ISSN electronic edition: 1617-9692
ISBN 978-3-540-74582-2 Springer Berlin Heidelberg New York
DOI 10.1007/978-3-540-74587-7

Springer is a part of Springer Science+Business Media
springer.com
© Springer-Verlag Berlin Heidelberg 2008

Typesetting by the author and SPi using a Springer LaTeX macro package

Cover design: *design & production* GmbH, Heidelberg

Printed on acid-free paper   SPIN: 12114160   41/SPi   5 4 3 2 1 0

I dedicate this book to my parents, James and Joyce Wilson.

Preface

Littlewood-Paley theory can be thought of as a profound generalization of the Pythagorean theorem. If $x \in \mathbf{R}^d$—say, $x = (x_1, x_2, \ldots, x_d)$—then we define x's norm, $\|x\|$, to be $(\sum_1^d x_n^2)^{1/2}$. This norm has the good property that, if $y = (y_1, y_2, \ldots, y_d)$ is any other vector in $\mathbf{R}^d$, and $|y_n| \leq |x_n|$ for each n, then $\|y\| \leq \|x\|$. In other words, the size of x, as measured by the norm function, is determined entirely by the sizes of x's components. This remains true if we let the dimension d increase to infinity, and define the norm of a vector (actually, an infinite sequence) $x = (x_1, x_2, \ldots)$ to be $\|x\| \equiv (\sum_1^\infty x_n^2)^{1/2}$.

In analysis it is often convenient (and indispensable) to decompose functions f into infinite series,

$$f(x) = \sum \lambda_n \phi_n(x), \tag{0.1}$$

where the functions ϕ_n come from some standard family (such as the Fourier system) and the λ_n's are complex numbers. (For the time being we will not specify how the series 0.1 is supposed to converge.) Typically the coefficients λ_n are defined by integrals of f against some other functions ψ_n. If we are interested about convergence in the sense of L^2 (or "mean-square"), and if the ϕ_n's comprise a *complete orthonormal family*, then each ψ_n can be taken to be $\bar{\phi}_n$, the complex conjugate of ϕ_n; i.e.,

$$\lambda_n = \int f(x)\, \bar{\phi}_n(x)\, dx,$$

and we have

$$\int |f(x)|^2\, dx = \sum |\lambda_n|^2.$$

(For the time being we will not specify the *domain* on which f and the ϕ_n's are defined.) If we are only interested in L^2 functions, then the natural norm,

$$\|f\|_2 \equiv \left(\int |f(x)|^2\, dx \right)^{1/2} = \left(\sum |\lambda_n|^2 \right)^{1/2},$$

has the same domination property possessed by the Euclidean norm on $\mathbf{R}^d$: if $g = \sum \gamma_n \phi_n$ and $|\gamma_n| \le |\lambda_n|$ for all n, then $\|g\|_2 \le \|f\|_2$. Even better, if, for some $\epsilon > 0$, we have $|\gamma_n| \le \epsilon |\lambda_n|$ for all n, then $\|g\|_2 \le \epsilon \|f\|_2$.

Unfortunately, L^2 is not always the most useful function space for a given problem. We might want to work in L^4, with its norm defined by

$$\|f\|_4 \equiv \left(\int |f(x)|^4 \, dx \right)^{1/4}.$$

To make things specific, let's suppose that our functions are defined on $[0, 1)$. The collection $\{\exp(2\pi inx)\}_{-\infty}^{\infty}$ defines a complete orthonormal family in $L^2[0, 1)$. Now, if $f \in L^4[0, 1)$, then the coefficients

$$\lambda_n \equiv \int_0^1 f(x) \, \exp(-2\pi inx) \, dx$$

are defined, and the infinite series,

$$\sum_{-\infty}^{\infty} \lambda_n \exp(2\pi inx),$$

converges to f in the L^4 sense, if we sum it up right. But the domination property fails in a very strong sense. Given $f \in L^4$, and given an integrable function g such $|\gamma_n| \le |\lambda_n|$ for all n, where

$$\gamma_n = \int_0^1 g(x) \, \exp(-2\pi inx) \, dx,$$

there is no reason to expect that $\|g\|_4$ is even finite, let alone controlled by $\|f\|_4$.

Littlewood-Paley theory provides a way to *almost* preserve the domination property. To each function f, one associates something called the *square function of* f, denoted $S(f)$. (Actually, the square function comes in many guises, but we will not go into that now.) Each square function is defined via inner products with a fixed collection of functions. Sometimes this collection is a complete orthonormal family for L^2, but it doesn't have to be. The square function $S(f)(x)$ is defined as a weighted sum (or integral) of the squares of the inner products, $|\langle f, \phi \rangle|^2$, where ϕ belongs to the fixed collection. The function $S(f)(x)$ varies from point to point, but, if f and g are two functions such that $|\langle g, \phi \rangle| \le |\langle f, \phi \rangle|$ for all ϕ, then $S(g)(x) \le S(f)(x)$ everywhere. The square function $S(f)$ also has the property that, if $1 < p < \infty$, and $f \in L^p$, the L^p norms of $S(f)$ and f are *comparable*.

The combination of these two facts—domination plus comparablility—lets us, in many situations, reduce the analysis of infinite series of functions,

$$f(x) = \sum \lambda_i \phi_i(x),$$

to the analysis of infinite series of *non-negative* functions,

$$S(f)(x) = \left(\sum |\gamma_i|^2 |\psi_i(x)|^2 \right)^{1/2} ;$$

and that greatly simplifies things. We have already mentioned the practice, common in analysis, of cutting a function into infinitely many pieces. Typically we do this to solve a problem, such as a PDE. We break the data into infinitely many pieces, solve the problem on each piece, and then sum the "piece-wise" solutions. The sums encountered this way are likely to contain a lot of complicated cancelations. Littlewood-Paley theory lets us control them by means of sums that have *no* cancelations.

The mutual control between $|f|$ and $S(f)$ is very tight. We will soon show that, if f is a bounded function defined on $[0, 1)$, there is a positive α such that $\exp(\alpha(S(f))^2)$ is integrable on $[0, 1)$—*and vice versa*. (This is not quite like saying that $|f|$ and $S(f)$ are pointwise comparable, but in many applications they might as well be.) This tight control is expressed quantitatively in terms of weighted norm inequalities. The reader will learn some sufficient (and not terribly restrictive) conditions on pairs of weights which ensure that

$$\int |f(x)|^p \, v \, dx \leq \int (S(f)(x))^p \, w \, dx \qquad (0.2)$$

or

$$\int (S(f)(x))^p \, v \, dx \leq \int |f(x)|^p \, w \, dx \qquad (0.3)$$

holds for all f in suitable test classes, for various ranges of p (usually, $1 < p < \infty$). He will also learn some necessary conditions for such inequalities.

The usefulness of the square function (in its many guises) comes chiefly from the fact that, for many linear operators T, $S(T(f))$, the square function of $T(f)$, is bounded pointwise by a function $\tilde{S}(f)$, where $\tilde{S}(\cdot)$ is an operator similar to—and satisfying estimates similar to—$S(\cdot)$. This makes it possible to understand the behavior of T, because one can say: $|T(f)|$ is controlled by $S(T(f))$, which is controlled by $\tilde{S}(f)$, which is controlled by $|f|$. Obviously, the closer the connection between $|f|$ and $S(f)$, the more efficient this process will be. The exponential-square results (and the corresponding weighted norm inequalities) imply that this connection is pretty close.

We have tried to make this book self-contained, not too long, and accessible to non-experts. We have also tried to avoid excessive overlap with other books on weighted norm inequalities. Therefore we have not treated every topic of relevance to weighted Littlewood-Paley theory. We have not touched on multi-parameter analysis at all, and we have dealt only briefly with vector-valued inequalities. We discuss A_p weights mainly with reference to the square function and singular integral operators. We prove the boundedness of the Hardy-Littlewood operator on $L^p(w)$ for $w \in A_p$ and we prove an extrapolation result—because we need both—but we don't prove A_p factorization or the Rubio de Francia extrapolation theorem, excellent treatments of which can be found in many books (e.g., [16] and [24]).

The book is laid out this way. Chapter 1 covers some basic facts from harmonic analysis. Most of the material there will be review for many people, but we have tried to present it so as not to intimidate the non-experts. Chapter 2 introduces the one-dimensional dyadic square function and proves some of its properties; it also introduces a few more techniques from harmonic analysis. In chapter 3 we prove the exponential-square estimates mentioned above (in one dimension only). These lead to an in-depth look at weighted norm inequalities. In chapter 4 we extend the results of the preceding chapters to d dimensions and to continuous analogues of the dyadic square function.

Chapters 5, 6, and 7 are devoted to the Calderón reproducing formula. The Calderón formula provides a canonical way of expressing "arbitrary" functions as linear sums of special, smooth, compactly supported functions. It is the foundation of wavelet theory. Aside from some casual remarks[1], we don't talk about wavelets. The expert will see the close connections between wavelets and the material in chapters 5–7. The non-expert doesn't have to worry about them to understand the material; but, should he ever encounter wavelets, a good grasp of the Calderón formula will come in very handy. We have devoted three chapters to it because we believe the reader will gain more by seeing essentially the same problem (the convergence of the Calderón integral formula) treated in increasing levels of generality, than in having one big portmanteau theorem dumped onto his lap. The portmanteau theorem (Theorem 7.1) does come; but we trust that, when it does, the reader is more than able to bear its weight.

Chapters 8 and 9 give some straightforward applications of weighted Littlewood-Paley theory to the analysis of Schrödinger and singular integral operators. This material could easily have come after that in chapter 10, but we felt that, where it is, it gave the reader a well-earned break from purely theoretical discussions.

In chapter 10 we return to theory. The scale of Orlicz spaces (which includes that of L^p spaces for $1 \leq p \leq \infty$) provides a flexible way of keeping track of the integrability properties of functions. It is very useful in the study of weighted norm inequalities. The material here *could* have come at the very beginning, but we felt that the reader would understand this theory better if he first saw the need for it.

As an application of Orlicz space theory, chapter 11 presents a different proof of Theorem 3.8 from chapter 3. This ingenious argument, due to Fedor Nazarov, completely avoids the use of good-λ inequalities (which we introduce in chapter 2). These have been a mainstay of analysis since the early 1970s. In the opinion of some researchers, they have also become a crutch. We are neutral on this issue, but please see our note at the end of chapter 2.

Chapter 12 applies the theory from the preceding chapters to give a new (and, we hope, accessible) proof of the Hörmander-Mihlin multiplier theorem. Chapter 13 extends the main weighted norm results from earlier chapters

---

[1] Like this one.

to the ℓ^2-valued setting. In chapter 14 we prove one theorem (Khinchin's Inequalities), but our discussion there is mainly philosophical. We look at what Littlewood-Paley theory can tell us about *pointwise* summation errors of Haar function expansions.

We have put exercises at the end of almost every chapter. Some of them expand on topics treated in the text; some tie up loose ends in proofs; some are referred to later in the book. We encourage the reader to understand all of them and to attempt at least a few. (We have supplied hints for the more difficult ones.)

The author wishes to thank the many colleagues who have offered suggestions, helped him track down references, and steered him away from blunders. These colleagues include David Cruz-Uribe, SFO (of Trinity University in Hartford, Connecticut), Doug Kurtz (of New Mexico State University), José Martell (of the Universidad Autónoma de Madrid), Fedor Nazarov (of Michigan State University), Carlos Pérez Moreno (of the Universidad de Sevilla), and Richard Wheeden (of Rutgers University, New Brunswick). I must particularly thank Roger Cooke, now retired from the University of Vermont, who read early drafts of the first chapters, and whose insightful criticisms have made them much more intelligible and digestible.

The author could not have written this book without the generous support of the Spanish Ministerio de Educación, Cultura, y Deporte, which provided him with a research grant (SAB2003-0003) during his 2004-2005 sabbatical at the Universidad de Sevilla. My family and I are indebted to so many members of the Facultad de Matematicas for their hospitality, that I hesitate to try to name them, for fear of omitting some. However, I must especially point out the kindness of my friend and colleague, Carlos Pérez Moreno. Without his tireless efforts, our visit to Sevilla would never have taken place. I do not have adequate words to express how much my family and I owe to him for everything he did for us, both before and after we arrived in Spain. Carlos, Sevilla, y España se quedarán siempre en nuestros corazones.

Contents

1

Some Assumptions

Every area of mathematics—and, indeed, of learning—is a minefield of presuppositions and "well-known" results. In this section I will try to acquaint the reader with what things are taken for granted in weighted Littlewood-Paley theory. Some of these things are definitions and notations, and some of them are theorems.

We assume that the reader has had a graduate course in measure theory, at least to the level of chapters 5 and 6 in [21]. Roughly speaking, this includes: the theory of the Lebesgue integral (in 1 and d dimensions), L^p spaces in $\mathbf{R}^d$ ($1 \leq p \leq \infty$) and their duals, and some functional analysis. We also assume that the reader knows a little about the Fourier transform.

We will use certain definitions and conventions repeatedly.

The definition of the Fourier transform we shall adopt is:

$$\hat{f}(\xi) \equiv \int_{\mathbf{R}^d} f(x)\, e^{-2\pi i x \cdot \xi}\, dx,$$

originally defined for $f \in L^1(\mathbf{R}^d)$, and then by extension to $f \in L^2$. We have the Fourier inversion formula

$$f(x) = \int_{\mathbf{R}^d} \hat{f}(\xi)\, e^{2\pi i x \cdot \xi}\, d\xi,$$

which holds pointwise for appropriate f, and extends, by beginning functional analysis, to all $f \in L^2$.

Our definition of the Fourier transform satisfies

$$\|f\|_2 = \|\hat{f}\|_2$$

and

$$(\widehat{f * g})(\xi) = \hat{f}(\xi)\hat{g}(\xi),$$

where $f * g$ is the usual convolution,

$$f * g(x) = \int_{\mathbf{R}^d} f(x - y)\, g(y)\, dy = \int_{\mathbf{R}^d} f(y)\, g(x - y)\, dy$$

defined for appropriate pairs of functions f and g.

We use $\mathcal{C}_0^\infty(\mathbf{R}^d)$ to denote the family of infinitely differentiable functions with compact supports. The Schwartz class $\mathcal{S}(\mathbf{R}^d)$ is the family of infinitely differentiable functions f such that, for all differential monomials D^α and all positive integers M,

$$\int_{\mathbf{R}^d} |D^\alpha f|\, (1 + |x|)^M\, dx < \infty.$$

A measurable function f is said to be *locally integrable* if $\int_K |f|\, dx < \infty$ for every compact subset of f's domain. This domain will always be $\mathbf{R}^d$ or some nice subset of it (such as an interval, ball, rectangle, or half-space). The only half-space we ever look at is $\mathbf{R}_+^{d+1}$, which equals $\mathbf{R}^d \times (0, \infty)$. We denote the space of locally integrable functions defined on $\mathbf{R}^d$ by $L^1_{\text{loc}}(\mathbf{R}^d)$.

If E is a measurable subset of $\mathbf{R}^d$, we denote E's Lebesgue measure by $|E|$. We will try to make it clear from the context when $|\cdot|$ means the measure of a set and when it means the absolute value of a number. We will also use $|\cdot|$ to denote the norm of a vector in $\mathbf{R}^d$.

If $I \subset \mathbf{R}$ is an interval, we let $\ell(I)$ denote I's length (which is the same as $|I|$). A *cube* $Q \subset \mathbf{R}^d$ is a cartesian product of d intervals all having the same length. We refer to this common length by $\ell(Q)$, and we call it Q's *sidelength*. Notice that $|Q| = \ell(Q)^d$.

Incidentally, we use '$\subset$' to denote "subset," not just "proper subset."

A dyadic interval is one of the form $[j/2^k, (j + 1)/2^k)$, where j and k are integers. A dyadic cube $Q \subset \mathbf{R}^d$ is a cube whose component intervals are all dyadic. The family of all dyadic cubes in $\mathbf{R}^d$ is denoted by $\mathcal{D}_d$. Strictly speaking, the family of dyadic intervals should be $\mathcal{D}_1$, but we will usually refer to it by $\mathcal{D}$.

The reader's first exercise is to show that, if Q and Q' are two dyadic cubes in $\mathbf{R}^d$, then either $Q \subset Q'$, $Q' \subset Q$, or $Q \cap Q' = \emptyset$.

The reader's second exercise is to show that, if $I \subset \mathbf{R}$ is any bounded interval, there exist dyadic intervals I_1, I_2, and I_3, all having the same length 2^{-k}, such that $(1/2)\ell(I) < 2^{-k} \leq \ell(I)$ and $I \subset I_1 \cup I_2 \cup I_3$. (This is like saying that every interval is "almost" a dyadic interval.)

The reader's third exercise is to generalize the second exercise to $\mathbf{R}^d$.

The first exercise has this consequence: If $\mathcal{F} \subset \mathcal{D}_d$ is any collection of dyadic cubes such that

$$\sup_{Q \in \mathcal{F}} \ell(Q) < \infty \tag{1.1}$$

then there exists a *disjoint* collection $\mathcal{F}' \subset \mathcal{F}$ such that every $Q \in \mathcal{F}$ is contained in some $Q' \in \mathcal{F}'$. The proof is: For every $Q \in \mathcal{F}$, let Q' be the *maximal* element (in the sense of set inclusion) of $\mathcal{F}$ that contains Q; such a maximal element must exist because of 1.1. The collection of all such Q''s is $\mathcal{F}'$.

If f is locally integrable and E is an appropriate measurable subset of f's domain (in practice, E is almost always a cube), then f_E is f's average value over E, defined by

$$f_E \equiv \frac{1}{|E|} \int_E f \, dx.$$

Very early in the book, the reader will encounter two conventions, endemic in Fourier analysis, which he might find a little disturbing. They are "the constantly changing constant"[1] and the use of '$\sim$'.

Much of analysis is about proving inequalities. We have two positive quantities—call them A and B—that depend on something else: a variable, a vector, a function, or some combination of these. Suppose it's a variable t. We typically want to show that there is a positive, finite constant C so that, for all t under consideration,

$$A(t) \leq C B(t). \tag{1.2}$$

We often want to prove such inequalities because they help us prove *equations*. For example, we might have two complicated but continuous functions $f(t)$ and $g(t)$, and want to show $f(0) = g(0)$. This is an immediate consequence of:

$$|f(t) - g(t)| \leq C B(t)$$
$$\lim_{t \to 0} B(t) = 0.$$

In practice, $|f(t) - g(t)|$ is hard to estimate directly, but $B(t)$ is easy (or easier) to handle.

An inequality like 1.2 usually follows from a chain of inequalities, like so,

$$\begin{aligned} A(t) &\leq C_1 A_1(t) \\ &\leq C_2 A_2(t) \\ &\leq \cdots \\ &\leq C_{129} B(t), \end{aligned}$$

but usually not so long. However, unless one is keeping careful track of the constants C_k, it is normal for analysts to write the preceding inequalities as:

$$\begin{aligned} A(t) &\leq C A_1(t) \\ &\leq C A_2(t) \\ &\leq \cdots \\ &\leq C B(t). \end{aligned}$$

The "constant" C, which is understood to be different at every stage, is "the constantly changing constant." It is understood that C can depend on the functions A_1, etc., but NOT on the parameter t.

---

[1] The name is due to T. W. Körner.

The other disturbing convention also concerns inequalities. Sticking with our example, suppose there are two positive, finite constants, C_1 and C_2, such that, for all relevant t,

$$C_1 B(t) \leq A(t) \leq C_2 B(t). \tag{1.3}$$

The inequalities 1.3 say that A and B are roughly the same size. An example of such a pair of functions is $A(t) = t(\log(e + t))$ and $B(t) = t(\log(534 + t^{25}))$, where the range of admissible t's is $[0, \infty)$; the reader should check this. A more interesting example is given by $A(t) = t/(\log(e + t))^2$ and $B(t) =$ the inverse function of $t(\log(e + t))^2$—and the reader should check this one, too.

Unless we are very interested in the values of C_1 and C_2, we will often express relationships like 1.3 by means of the "semi-equation,"

$$A(t) \sim B(t).$$

When the context does not make it clear, we will say what the admissible t's are.

We will conclude this section with a deceptively simple observation and a few of its profound consequences.

Suppose that f is a locally integrable function with the property that, for every $\epsilon > 0$, there exists an $R > 0$ such that, if Q is any cube with $\ell(Q) > R$, then

$$\frac{1}{|Q|} \int_Q |f| \, dx < \epsilon. \tag{1.4}$$

This hypothesis is not very restrictive: it is satisfied by every $f \in L^p$, for all $1 \leq p < \infty$ (but not for $p = \infty$). Let λ be a positive number, and let $\mathcal{F}_\lambda$ be the family of dyadic cubes Q such that

$$\frac{1}{|Q|} \int_Q |f| \, dx > \lambda.$$

Our hypothesis on f implies that every $Q \in \mathcal{F}_\lambda$ is contained in some maximal $Q' \in \mathcal{F}_\lambda$. (This, by the way, holds even if $\mathcal{F}_\lambda$ is empty: check the logic!)

Call this family of maximal cubes $\mathcal{F}_\lambda'$.

If $Q \in \mathcal{F}_\lambda'$, then

$$\frac{1}{|Q|} \int_Q |f| \, dx > \lambda.$$

I claim that, as well,

$$\frac{1}{|Q|} \int_Q |f| \, dx \leq 2^d \lambda.$$

To see this, let $\tilde{Q}$ be the unique dyadic cube such that $Q \subset \tilde{Q}$ and $\ell(\tilde{Q}) = 2\ell(Q)$. Then, because of Q's maximality,

$$\frac{1}{|\tilde{Q}|} \int_{\tilde{Q}} |f| \, dx \leq \lambda.$$

But

$$\frac{1}{|Q|} \int_Q |f|\, dx \le \frac{1}{|Q|} \int_{\tilde{Q}} |f|\, dx \le \frac{|\tilde{Q}|}{|Q|} \lambda,$$

from which the inequality follows.

Notice that, by the Lebesgue differentiation theorem, $|f| \le \lambda$ almost everywhere off the set $\cup_{\mathcal{F}'_\lambda} Q$.

Now, what is this good for?

Harmonic analysis is about the action of linear operators on functions. Usually we are trying to show that some operator T is bounded on some L^p; i.e., we wish to show that there is a constant A such that, for all $f \in L^p$,

$$\|T(f)\|_p \le A\|f\|_p.$$

This is often accomplished by *splitting the function f* into finitely many pieces,

$$f = \sum_1^N f_i,$$

and showing that each $T(f_i)$ is "well-behaved" in some fashion. (There are also times when we split f into infinitely many pieces, but that story can wait.) Such splittings typically work because the pieces f_i are "well-behaved"—and so give rise to good $T(f_i)$'s—for different reasons.

The best-known and most widely used technique for splitting functions, due to A. P. Calderón and A. Zygmund, is based on the foregoing observation about dyadic cubes. It lets us write any function f satisfying 1.4 as the sum of two functions (usually called g and b, for "good" and "bad"). As we shall see, "good" and "bad" must be used advisedly, because the functions g and b are both "good" (also "bad"), but in different ways.

Theorem 1.1. *Let f satisfy 1.4. For every $\lambda > 0$, there is a (possibly empty) family $\mathcal{F}$ of pairwise disjoint dyadic cubes such that $f = g + b$, where $\|g\|_\infty \le 2^d \lambda$ and $b = \sum_{Q \in \mathcal{F}} b_{(Q)}$. Each function $b_{(Q)}$ has its support contained in Q and satisfies*

$$\int b_{(Q)}\, dx = 0$$

and

$$\int |b_{(Q)}|\, dx \le 2^d \lambda |Q|.$$

Moreover, the family $\mathcal{F}$ can be chosen so that

$$\sum_{\mathcal{F}} |Q| \le \frac{2}{\lambda} \int_{|f| > \lambda/2} |f|\, dx. \tag{1.5}$$

Before proving the theorem, we should explain how the functions g and b are good and bad in their own ways.

The function g is good because it is bounded. It is bad because it might have unbounded support.

The function b is bad because it is in general unbounded. However, it is good because it is a sum of non-interfering pieces (disjoint supports) with controlled L^1 norms, and which satisfy a cancelation condition, and with a total support that is also controlled, at least in terms of measure.

Proof of Theorem 1.1. We will essentially prove the theorem twice. The first proof will give us g and b that almost do what we want. Then we will show how, with only a small modification, we can get the desired g and b.

To begin: let $\mathcal{F}_\lambda$ (note that we have dropped the 'prime') be the family of maximal dyadic cubes satisfying

$$\frac{1}{|Q|} \int_Q |f| \, dx > \lambda.$$

By our observation,

$$\frac{1}{|Q|} \int_Q |f| \, dx \leq 2^d \lambda$$

for every $Q \in \mathcal{F}_\lambda$.

Define

$$g(x) = \begin{cases} f(x) & \text{if } x \notin \cup_{\mathcal{F}_\lambda} Q; \\ \frac{1}{|Q|} \int_Q f \, dt & \text{if } x \in Q \in \mathcal{F}_\lambda. \end{cases}$$

Then g is clearly bounded by $2^d \lambda$ almost everywhere. Set $b = f - g$. A little computation shows

$$b(x) = \begin{cases} 0 & x \notin \cup_{\mathcal{F}_\lambda} Q; \\ f(x) - f_Q & \text{if } x \in Q \in \mathcal{F}_\lambda. \end{cases}$$

We set $b_{(Q)}(x) = (f(x) - f_Q)\chi_Q(x)$. Then this gives the desired splitting in most respects. We have the right bound on g and, for every $Q \in \mathcal{F}_\lambda$,

$$\int |b_{(Q)}| \, dx \leq 2^{d+1} \lambda |Q|,$$

which is only off by a factor of 2. What about 1.5? I claim that we are close to having it. Every $Q \in \mathcal{F}_\lambda$ satisfies

$$|Q| \leq \frac{1}{\lambda} \int_Q |f| \, dt.$$

But these cubes are also disjoint. Therefore

$$\sum_{\mathcal{F}_\lambda} |Q| \leq \sum_{\mathcal{F}_\lambda} \frac{1}{\lambda} \int_Q |f| \, dt \leq \frac{1}{\lambda} \int |f| \, dt,$$

which is nearly right.

To get our final decomposition, we first split f in a naive fashion. Set

$$f_1(x) = \begin{cases} f(x) & \text{if } |f(x)| > \lambda/2; \\ 0 & \text{otherwise,} \end{cases}$$

and define $f_2 = f - f_1$. Notice that $|f_2| \leq \lambda/2$ everywhere. Now we apply the previous splitting argument to f_1, but use $\lambda/2$ as our cut-off height, instead of λ. We obtain two functions $\tilde{g}$ and $\tilde{b}$, and a disjoint family of dyadic cubes $\tilde{\mathcal{F}}_\lambda$ such that $\tilde{b} = \sum_{\tilde{\mathcal{F}}_\lambda} \tilde{b}_{(Q)}$, where the functions $\tilde{b}_{(Q)}$ satisfy the support and cancelation conditions, and also have

$$\int |\tilde{b}_{(Q)}| \, dx \leq 2^{d+1}(\lambda/2)|Q| = 2^d \lambda |Q|.$$

Summing up the measures of the Q's, we get

$$\sum_{\tilde{\mathcal{F}}_\lambda} |Q| \leq \sum_{\tilde{\mathcal{F}}_\lambda} \frac{2}{\lambda} \int_Q |f_1| \, dt \leq \frac{2}{\lambda} \int_{|f|>\lambda/2} |f| \, dt,$$

which is 1.5.

We now set $b = \tilde{b}$ and $g = \tilde{g} + f_2$. Then b is what we want, and $|g| \leq \lambda/2 + 2^d \lambda/2 \leq 2^d \lambda$. That finishes the proof.

The reader could reasonably ask what purpose is served by being able to split f this way for every positive λ. Why isn't it enough to split it just for $\lambda = 1$? That is another story that will have to wait.

Notes

The Calderón-Zygmund decomposition (Theorem 1.1) first appears in [8]. The treatment here is based on that in [53].

An Elementary Introduction

We begin with the simplest object in Littlewood-Paley theory: the dyadic square function.

Let $\mathcal{D}$ be the collection of dyadic intervals on $\mathbf{R}$. For every $I \in \mathcal{D}$, we let I_l and I_r denote (respectively) the left and right halves of I. For each $I \in \mathcal{D}$, set

$$h_{(I)}(x) \equiv \begin{cases} |I|^{-1/2} & \text{if } x \in I_l; \\ -|I|^{-1/2} & \text{if } x \in I_r; \\ 0 & \text{if } x \notin I. \end{cases}$$

Notice that each $h_{(I)}$ satisfies $\int h_{(I)} = 0$ and $\|h_{(I)}\|_2 = 1$. These functions $h_{(I)}$ are known as the *Haar functions*.

We claim that $\{h_{(I)}\}_{I \in \mathcal{D}}$ is an orthonormal system for $L^2(\mathbf{R})$. We've just seen that

$$\int h_{(I)}(x)\, h_{(I)}(x)\, dx = 1.$$

Suppose I and J belong to $\mathcal{D}$ and $I \neq J$. If $I \cap J = \emptyset$ it is trivial that

$$\int h_{(I)}(x)\, h_{(J)}(x)\, dx = 0.$$

Suppose $I \cap J \neq \emptyset$. Without loss of generality, we may assume that $I \subset J$. But then, since $I \neq J$, the support of $h_{(I)}$ is entirely contained in J_l or J_r—across which $h_{(J)}$ is constant. But $\int h_{(I)} = 0$, and so

$$\int h_{(I)}(x)\, h_{(J)}(x)\, dx = 0$$

in this case, too.

For any $f \in L^1_{loc}(\mathbf{R})$ and $I \in \mathcal{D}$, we define

$$\begin{aligned} \lambda_I(f) &= \int f(x)\, h_{(I)}(x)\, dx \\ &= \langle f, h_{(I)} \rangle, \end{aligned}$$

where we are using $\langle \cdot, \cdot \rangle$ to denote the usual L^2 inner product. The number $\lambda_I(f)$ is known as f's *Haar coefficient* for the interval I. By Bessel's Inequality, we immediately have:

$$\sum_I |\lambda_I(f)|^2 \leq \int |f(x)|^2 \, dx$$

for any $f \in L^2(\mathbf{R})$. We claim that the Haar functions actually form a complete orthonormal system for $L^2(\mathbf{R})$. To see this, take $f \in L^2$, and suppose that $\langle f \, h_{(I)} \rangle = 0$ for all $I \in \mathcal{D}$. The claim will be proved if we can show that $f = 0$. For this it is sufficient to prove that f is (a.e.) constant on $(-\infty, 0)$ and $(0, \infty)$.

A little computation shows that, for any $I \in \mathcal{D}$,

$$\lambda_I(f) = (f_{I_l} - f_I)|I|^{1/2} = -(f_{I_r} - f_I)|I|^{1/2}, \tag{2.1}$$

where we are using the convention that f_J equals f's average over the interval J: $f_J = \frac{1}{|J|} \int_J f$. Now, let x and y belong to $(0, \infty)$. There is a minimal dyadic interval I_0 such that both x and y belong to I_0. Suppose that J_0 and K_0 are two very small dyadic intervals such that

$$x \in J_0 \subset I_0$$
$$y \in K_0 \subset I_0.$$

We can find a sequence of nested dyadic intervals $J_0 \subset J_1 \subset J_2 \subset \ldots \subset J_n \equiv I_0$ so that each J_k is a right or left half of J_{k+1}. Because of 2.1, we must have

$$f_{J_0} = f_{J_1} = \cdots = f_{J_n} = f_{I_0}.$$

But a similar argument proves that $f_{K_0} = f_{I_0}$. Thus $f_{J_0} = f_{K_0}$. Since J_0 and K_0 are arbitrarily small, the Lebesgue differentiation theorem implies that f is a.e. constant on $(0, \infty)$. Obviously, the same argument works as well on $(-\infty, 0)$. This proves completeness.

Elementary functional analysis now implies that, for all $f \in L^2(\mathbf{R})$, the sum $\sum_I \lambda_I(f)h_{(I)}$ converges to f in L^2, and that

$$\int |f(x)|^2 \, dx = \sum_I |\lambda_I(f)|^2. \tag{2.2}$$

We will be saying a lot about 2.2, but before we start that, it might be a good idea to say a few words about the convergence of $\sum_I \lambda_I(f)h_{(I)}$. This is not just any old Hilbert space sum.

Suppose that f's support is contained inside $[0, 1)$ and that $\int f \, dx = 0$. The reader should satisfy himself of the truth of the following statement: If $I \in \mathcal{D}$ and $I \not\subset [0, 1)$, then $\lambda_I(f) = 0$. This says that the sum $\sum_I \lambda_I(f)h_{(I)}$ is, in a very useful sense, *localized*. For, suppose now that f is an arbitrary locally integrable function. Write $f = f_1 + f_2$, where

$$f_1(x) = \begin{cases} f(x) - \int_0^1 f\, dt & \text{if } x \in [0,1); \\ 0 & \text{otherwise.} \end{cases}$$

Again, the reader should see for himself that

$$\lambda_I(f) = \begin{cases} \lambda_I(f_1) & \text{if } I \subset [0,1); \\ \lambda_I(f_2) & \text{otherwise.} \end{cases}$$

and that, in fact, if J is any dyadic interval, and we split f into $f_1 + f_2$, where

$$f_1(x) = \begin{cases} f(x) - f_J & \text{if } x \in J; \\ 0 & \text{otherwise,} \end{cases}$$

we will have

$$\lambda_I(f) = \begin{cases} \lambda_I(f_1) & \text{if } I \subset J; \\ \lambda_I(f_2) & \text{otherwise.} \end{cases}$$

We can even go further. Suppose that $\{J_k\}_k$ is an arbitrary *disjoint* collection of dyadic intervals. Now split f into $f_1 + f_2$, where

$$f_1(x) = \begin{cases} f(x) - f_{J_k} & \text{if } x \in J_k; \\ 0 & \text{if } x \notin \cup_k J_k. \end{cases} \tag{2.3}$$

This definition forces f_2 to equal

$$f_2(x) = \begin{cases} f_{J_k} & \text{if } x \in J_k; \\ f(x) & \text{if } x \notin \cup_k J_k. \end{cases}$$

This splitting has the consequence that, if I is any dyadic interval not *properly* contained in some J_k, then

$$\int_I f\, dx = \int_I f_2\, dx. \tag{2.4}$$

Establishing 2.4 is an *excellent* exercise for the reader. Equation 2.4 implies that, for all dyadic intervals I,

$$\lambda_I(f) = \begin{cases} \lambda_I(f_1) & \text{if } I \text{ is a subset of some } J_k; \\ \lambda_I(f_2) & \text{otherwise.} \end{cases} \tag{2.5}$$

To put all this in plain, but approximate, language: $\lambda_I(f)$ measures f's deviation from its mean, at the scale $(1/2)\ell(I)$, on I. This fact is expressed more precisely by 2.1. Another way to understand 2.5 is to think of the inner products $\langle f, h_{(I)} \rangle$ as filters that "catch" the action of f on I, at scale roughly equal to $\ell(I)$, and allow everything else to pass through.

If we combine 2.1 with the formula for $h_{(I)}$, we get this convenient fact: If I is any dyadic interval, and I' is I's right or left half, then

$$f_{I'} - f_I = \lambda_I(f) h_{(I)}(x),$$

where x is any point in I'.

Why is it convenient? Consider a "tower" of dyadic intervals $I_0 \subset I_1 \subset I_2 \subset \cdots \subset I_N$, where $\ell(I_{k+1}) = 2\ell(I_k)$ for all $0 \le k < N$. Then, for any $x \in I_0$:

$$f_{I_0} - f_{I_N} = \sum_0^{N-1} (f_{I_k} - f_{I_{k+1}})$$

$$= \sum_1^N \lambda_{I_k}(f) h_{(I_k)}(x).$$

We can rewrite the last sum as

$$\sum_{\substack{I \in \mathcal{D}:\, I_0 \subset I \\ \ell(I_0) < \ell(I) \le 2^N \ell(I_0)}} \lambda_I(f) h_{(I)}(x).$$

There is clearly nothing special about I_0, I_N, or x in this argument. We are allowed to say the following: If I and J are dyadic intervals, $I \subset J$, and $x \in I$, then

$$f_I - f_J = \sum_{\substack{K \in \mathcal{D}:\, I \subset K \subset J \\ \ell(I) < \ell(K) \le \ell(J)}} \lambda_K(f) h_{(K)}(x). \qquad (2.6)$$

But I has a length—call it 2^p—and so does J—call it 2^r (where, of course, $r \ge p$). If we sum up 2.6 over all dyadic I with length 2^p and all dyadic J with length 2^r, we get

$$\sum_{I \in \mathcal{D}:\ell(I)=2^p} f_I \chi_I(x) - \sum_{J \in \mathcal{D}:\ell(J)=2^r} f_J \chi_J(x) = \sum_{\substack{K \in \mathcal{D} \\ 2^p < \ell(K) \le 2^r}} \lambda_K(f) h_{(K)}(x). \quad (2.7)$$

Look closely at what is going on in 2.7. The sum on the far left is what you get when you replace f by its average values over dyadic intervals of length 2^p. The sum contains infinitely many terms, but, for every x, at most one term will be non-zero. Similar comments apply to the second sum on the left. The sum on the right side of the equals sign also has infinitely many terms. However, for any x, at most $r - p$ of these terms will be non-zero.

Before going further, the reader should prove to himself that everything I have just said is true.

Notice what happens if we let $p \to -\infty$ and $r \to \infty$. If $f \in L^2$, the far-left sum converges to f almost everywhere, while the second sum on the left converges to zero everywhere; so, the left-hand side of 2.7 converges to f almost everywhere. Meanwhile, the right-hand side of 2.7 will converge to f in L^2. THEREFORE, the left-hand side of 2.7 *also* converges to f in L^2, and the right-hand side *also* converges to f almost-everywhere.

Here is a question: To what extent does this equivalence extend to other L^p spaces, and even to weighted L^p spaces? (We'll explain what weighted spaces are in a bit.) To put it more generally: When can we use a vector-space decomposition of a function, via Haar functions, to get information about the function's actual values—and vice-versa? This is one way to phrase the problem that Littlewood-Paley theory tries to address.

Let's get back to 2.2. We begin by rewriting it in a funny way. Notice that

$$1 = \frac{1}{|I|} \int \chi_I(x)\, dx.$$

Therefore,

$$\sum_I |\lambda_I(f)|^2 = \sum_I |\lambda_I(f)|^2 \frac{1}{|I|} \int \chi_I(x)\, dx$$

$$= \int \left(\sum_I \frac{|\lambda_I(f)|^2}{|I|} \chi_I(x) \right)\, dx.$$

The *one-dimensional dyadic square function*, which makes sense for any $f \in L^1_{loc}(\mathbf{R})$, is defined by the equation

$$S(f)(x) = \left(\sum_I \frac{|\lambda_I(f)|^2}{|I|} \chi_I(x) \right)^{1/2}. \tag{2.8}$$

We have approached this formula through an L^2 result 2.2, which says that

$$\|f\|_2 = \|S(f)\|_2 \tag{2.9}$$

for $f \in L^2$. Before going one step further, it will be profitable to reflect on the meaning of 2.9. It is a peculiar equation.

Let's first consider

$$\|f\|_2 \leq \|S(f)\|_2. \tag{2.10}$$

Notice that $|f|^2 = |\sum_I \lambda_I(f) h_{(I)}|^2$, while $(S(f))^2 = \sum_I |\lambda_I(f) h_{(I)}|^2$. We usually expect the square of a sum to be a lot bigger than a sum of squares, but inequality 2.10 says that, on the average, this isn't true for sums of Haar functions. The reason is that the sum $\sum_I \lambda_I(f) h_{(I)}$ has a lot of cancelation in it. It's remarkable that this cancelation should work out so neatly as to give 2.10. If it doesn't seem remarkable, that's only because we've been spoiled by a too-close familiarity with functional analysis.

Now let's consider

$$\|f\|_2 \geq \|S(f)\|_2. \tag{2.11}$$

This says that the square of the sum is not, on the average, much smaller than the sum of squares. Given what was said in the preceding paragraph, that doesn't seem like a big deal; but it is, precisely because of the sum's cancelation. Inequality 2.11 says that there isn't *too* much cancelation in $\sum_I \lambda_I(f) h_{(I)}$. It only seems obvious because of Bessel's inequality, which is a general fact about Hilbert spaces, and so hides the numerical nitty-gritty in this special case.

Another way to consider 2.9 is in terms of energy and amplitude. Think of $f = f(t)$, a function of time, as if it were a signal. Then $\|f\|_2$ gives one measure of f's average amplitude. The expression $\|S(f)\|_2^2$, which is a sum of squares of f's component pieces, provides a measure of the energy of the signal. Equation 2.9 says that there is a direct relation between a signal's amplitude and its energy. Later, when we consider weighted forms of 2.10, and their applications to the study of Schrödinger operators, we will make the connection between amplitudes and energies more explicit.

We have emphasized the non-obviousness of 2.9 because the burden of Littlewood-Paley theory is to extend 2.9 to settings where it is patently *not* obvious. These include L^p ($p \neq 2$) and so-called weighted spaces, in which the underlying measure is no longer the familiar Lebesgue one. For example, it turns out that, if $1 < p < \infty$, there are constants c_p and C_p so that, for all $f \in L^p(\mathbf{R})$,

$$c_p\|f\|_p \leq \|S(f)\|_p \leq C_p\|f\|_p.$$

This fact is so central to what we will be doing that it deserves to be stated in a theorem.

Theorem 2.1. *For all $1 < p < \infty$, there are constants c_p and C_p, depending only on p, so that, for all $f \in L^p(\mathbf{R})$,*

$$c_p\|f\|_p \leq \|S(f)\|_p \leq C_p\|f\|_p. \tag{2.12}$$

We will prove Theorem 2.1 shortly. However, before doing so, we wish to describe one possible application of an inequality like 2.12. It may help to convince the reader that Littlewood-Paley theory really is good for something.

It involves that notorious buzzword-of-the-hour, "wavelets." Now, Haar functions aren't wavelets, strictly speaking, but they're near enough for this example. Suppose we have a function f which belongs to some $L^p(\mathbf{R})$, with $1 < p < \infty$. It is fundamentally important to know to what extent the sum $\sum_I \lambda_I(f)h_{(I)}$ really represents f. This is not a stupid question; and its non-stupidity comes from the fact that the coefficients $\lambda_I(f)$ are bound to have errors in them. In other words, the sum we have to contend with isn't $\sum_I \lambda_I(f)h_{(I)}$, but $\sum_I \lambda_I(f)h_{(I)} + \sum_I \epsilon_I \lambda_I h_{(I)}$, where the numbers ϵ_I reflect the relative errors in our "measurements" of the λ_I's. Inequality 2.12 says that if the ϵ_I's all have small absolute values, then the relative error in the sum will also be small, as measured in various L^p spaces. This property is, alas, *not* shared by Fourier coefficients for $p \neq 2$.

But why do we care about L^p for $p \neq 2$, when L^2 is so nice? The reason is that having a "small" L^p norm means different things for different p. For small p, it means that f has good decay at infinity; for large p, it tells us that f doesn't ever "spike" too sharply. So, an inequality like 2.12 says that

$\sum_I \lambda_I(f)h_{(I)}$ can represent f pretty well, and that good estimates of c_p and C_p tell us how careful we have to be in computing the $\lambda_I(f)$'s, if we want this representation to be faithful.

Our first major goal is a proof of Theorem 2.1. There are at least two approaches we could take here. We could give a quick-and-dirty, relatively direct proof. Unfortunately, this proof does not generalize to small p $(0 < p \le 1)$ or to weighted settings. The other proof—the one we will give—is more involved, but we believe it better shows what is going on in Theorem 2.1. It also generalizes readily. Readers wishing to see the quick and dirty proof first will find it sketched in exercises at the end of the next chapter. We feel it fair to warn the reader that this "simpler" proof has complications of its own, including discussions of things which, at first, seem to have nothing to do with the square function.

Our proof of Theorem 2.1 will not work directly with f, but with f's averages.

Definition 2.1. *Let* $f : \mathbf{R} \mapsto \mathbf{R}$ *be locally integrable. The dyadic maximal function of* f, f^*, *is given by:*

$$f^*(x) \equiv \sup_{I:x \in I \in \mathcal{D}} |f_I|.$$

The dyadic Hardy-Littlewood maximal function of f, $M_d(f)$, *is defined by:*

$$M_d(f)(x) \equiv \sup_{I:x \in I \in \mathcal{D}} |f|_I.$$

We will be seeing a lot of maximal functions like these. Before actually proving Theorem 2.1, we will make a rather long digression to investigate some of their properties. This will also provide a chance to introduce the sorts of arguments we will frequently see in this book, and to review some of the fundamentals from chapter 1.

By the Lebesgue differentiation theorem, $|f(x)| \le f^*(x)$ almost everywhere.

Because $M_d(f)$ has the absolute value inside its defining integral, we always have $f^*(x) \le M_d(f)(x)$.

What makes $M_d(f)$ really useful is the following:

Theorem 2.2. *For all* $1 < p \le \infty$ *there is a constant* C_p, *depending only* p, *such that, for all* $f \in L^p$,

$$\|M_d(f)\|_p \le C_p\|f\|_p. \tag{2.13}$$

We will prove Theorem 2.2, and we will apply some ideas from the proof to investigate the "fine structure" of M_d. (These fine structure properties will be used in the next chapter.) Then we will continue with the proof of Theorem 2.1.

The proof of Theorem 2.2 is based on the following well-known equation: *For any $0 < p < \infty$,*

$$\int_X |f(x)|^p \, d\mu = p \int_0^\infty \lambda^{p-1} \mu(\{x : |f(x)| > \lambda\}) \, d\lambda, \qquad (2.14)$$

which is valid for any measure space $(X, \mathcal{M}, \mu)$. (Equation 2.14 is frequently stated to hold for σ-finite measure spaces, and proved by Fubini-Tonelli. If the reader tries to do this, he will find that the trickiest step comes in proving the measurability of certain sets in $X \times (0, \infty)$. If he's lucky, he'll stumble upon a proof for the general case.)

Now, take $g \in L^1$ and let $\lambda > 0$. The set $\{x : M_d(g)(x) > \lambda\}$ consists of

$$S_\lambda = \cup \{I \in \mathcal{D} : |g|_I > \lambda\}.$$

But, if I is big enough, $|g|_I \leq \lambda$ (because $g \in L^1$). So, we can replace the union defining S_λ by a union over a subcollection; namely, those $I \in \mathcal{D}$ such that $|g|_I > \lambda$ and are *maximal* (in the sense of set inclusion). Call this set of maximal intervals $\{I_i\}$. Because these intervals are dyadic, they are pairwise disjoint.

Each I_i satisfies

$$\frac{1}{|I_i|} \int_{I_i} |g| \, dt > \lambda;$$

implying

$$|I_i| \leq \frac{1}{\lambda} \int_{I_i} |g| \, dt.$$

Therefore,

$$\begin{aligned} |S_\lambda| &= \sum_i |I_i| \\ &\leq \frac{1}{\lambda} \sum_i \int_{I_i} |g| \, dt \\ &\leq \frac{1}{\lambda} \int |g| \, dt. \end{aligned}$$

This is the celebrated weak-type inequality for the Hardy-Littlewood maximal function.

Now take $f \in L^p$, with $1 < p < \infty$, and write $f = f_1 + f_2$, where

$$f_1(x) = \begin{cases} f(x) & \text{if } |f(x)| > \lambda/2; \\ 0 & \text{otherwise,} \end{cases}$$

where λ is an arbitrary positive number. Then $M_d(f) \leq M_d(f_1) + M_d(f_2) \leq M_d(f_1) + \lambda/2$ (because $|f_2| \leq \lambda/2$ everywhere). Therefore,

$$\{x : M_d(f)(x) > \lambda\} \subset \{x : M_d(f_1)(x) > \lambda/2\}.$$

The celebrated inequality implies that

$$|\{x : \ M_d(f_1)(x) > \lambda/2\}| \le \frac{2}{\lambda} \int |f_1| \, dt$$

$$= \frac{2}{\lambda} \int_{\{t: \ |f(t)|>\lambda/2\}} |f(t)| \, dt.$$

We plug in our estimate for $|\{x : \ M_d(f)(x) > \lambda\}|$ and apply 2.14:

$$\int (M_d(f))^p \, dx \le p \int_0^\infty \lambda^{p-1} \left[\frac{2}{\lambda} \int_{\{t: \ |f(t)|>\lambda/2\}} |f(t)| \, dt \right] d\lambda$$

$$= \int |f(t)| \left[\int_0^{2|f(t)|} 2p\lambda^{p-2} \, d\lambda \right] dt$$

$$= 2^p (\frac{p}{p-1}) \int |f(t)||f(t)|^{p-1} \, dt$$

$$= 2^p (\frac{p}{p-1}) \int |f(t)|^p \, dt.$$

We've used Fubini-Tonelli in the antepenultimate line.

This implies $\|M_d(f)\|_p \le C_p \|f\|_p$ for $1 < p < \infty$. Of course, the case of $p = \infty$ is trivial.

The kind of argument used to prove Theorem 2.2 is called *interpolation*. We showed—or could see directly—that M_d was "controlled" on L^1 and L^∞, and we used that to prove that it was actually *bounded* on L^p, for $1 < p < \infty$. A general principle is lurking here, which the reader might want to investigate for himself. When he gets tired of that, he can turn to the end of this chapter, where we state and prove a special case of the Marcinkiewicz Interpolation Theorem.

The "fine structure" of M_d referred to above concerns its action on L^1. This might seem pointless, since M_d is obviously *not* bounded on $L^1(\mathbf{R})$: if $f = \chi_{[-1,1]}$, then $M_d(f) \sim |x|^{-1}$ when $|x|$ is large. However, if f's support is contained in a dyadic interval I, we have two substitute results. The first one is

$$\int_I |f(x)| \log(e + |f(x)|/|f|_I) \, dx \sim \int_I M_d(f) \, dx. \qquad (2.15)$$

The second result is

$$\frac{1}{|I|} \int_I (M_d(f))^\beta \, dx \le C_\beta \left(\frac{1}{|I|} \int_I |f| \, dx \right)^\beta, \qquad (2.16)$$

valid for $0 < \beta < 1$.

Inequality 2.15 says that, if we restrict our attention to an interval, then $M_d(f)$ is integrable if and only f, in a precise sense, is a little better than integrable. Inequality 2.16 says—with the same restriction—that, if f is integrable, then $M_d(f)$ is *almost* integrable.

We will prove these inequalities when $f \geq 0$ and $|I| = |f|_I = 1$, leaving the general cases as exercises.

The splitting argument we used to prove Theorem 2.2 implies that, for every $\lambda > 1$,

$$|\{x \in I : \ M_d(f)(x) > \lambda\}| \leq \frac{C}{\lambda} \int_{|f|>\lambda/2} |f(t)| \, dt, \qquad (2.17)$$

with the bound of

$$|\{x \in I : \ M_d(f)(x) > \lambda\}| \leq 1 \qquad (2.18)$$

for $\lambda \leq 1$. If we multiply both sides of these estimates by $p\lambda^{p-1}$ (i.e., 1) and integrate from 0 to ∞, we get

$$\int M_d(f) \, dx \leq \int_0^1 |\{x \in I : \ M_d(f)(x) > \lambda\}| \, d\lambda$$

$$+ \int_1^\infty \left(\frac{C}{\lambda} \int_{|f|>\lambda/2} |f(t)| \, dt \right) d\lambda$$

$$\leq 1 + C \int_I |f(t)| \left(\int_1^{2|f(t)|} \frac{C}{\lambda} \, d\lambda \right)_+ dt$$

$$\leq 1 + \int_I |f(t)| \, \log^+(|f(t)|) \, dt$$

$$\leq C \int_I |f(x)| \, \log(e + |f(x)|) \, dx,$$

proving one direction of 2.15.

To show the other direction in 2.15, define $\{I_j^k\}_j$ ($k = 0, 1, 2, \ldots$) to be the family of maximal dyadic $I_j^k \subset I$ such that

$$f_{I_j^k} > 2^k.$$

By maximality, these intervals also satisfy

$$f_{I_j^k} \leq 2^{k+1}.$$

Then

$$M_d(f)(x) \sim 1 + \sum_{k,j} f_{I_j^k} \chi_{I_j^k}(x)$$

on I, since the sum on the right is comparable to $1 + \sum_0^\infty 2^k \chi_{\{M_d(f)>2^k\}}$. Therefore

$$\int_I M_d(f)(x) \, dx \sim 1 + \int \left(\sum_{k,j} f_{I_j^k} \chi_{I_j^k}(x) \right) dx$$

$$= 1 + \int f(x) \left(\sum_{k,j} \chi_{I_j^k}(x) \right) dx$$

$$\sim \int_I f(x) \, \log(e + M_d(f)(x)) \, dx$$

$$\geq C \int_I f(x) \, \log(e + f(x)) \, dx,$$

proving 2.15.

Inequality 2.16 is proved very much in the same manner as the second inequality in 2.15. Simply multiply the estimates 2.17 and 2.18 by $\beta\lambda^{\beta-1}$ and integrate from 0 to ∞. The integral from 0 to 1 is bounded by 1 (because of 2.18), and the integral from 1 to ∞ is bounded by a constant times

$$\int_I |f| \left(\int_1^{2|f|} \lambda^{\beta-2} \, d\lambda \right)_+ \, dx \leq C_\beta,$$

from our normalization on f and the fact that $\beta < 1$.

Note that 2.15 implies that $M_d(f)$ is not bounded on $L^1(I)$, and it even points the way to building a counterexample. For example, the function defined by

$$f(x) = \begin{cases} \frac{2^k}{k^2} & \text{if } x \in [2^{-k}, 2^{-k+1}), \, k \geq 1; \\ 0 & \text{otherwise,} \end{cases}$$

is in L^1, but $\int_0^1 |f| \log^+(|f|) \, dx = \infty$, as the reader should have no trouble showing. (He might have a little more trouble showing directly that $\int_0^1 M_d(f) \, dx = \infty$, but we think it will do him good to try.)

This concludes our digression on M_d.

The proof of Theorem 2.1 will use the method of "good-λ inequalities." Since we will be using this method several times, we'll state it here in a general theorem.

Theorem 2.3. *Let f and g be non-negative measurable functions defined on a measure space $(\Omega, \mathcal{M}, \mu)$. Suppose that, for every $\epsilon > 0$, there is a $\gamma > 0$ such that, for all $\lambda > 0$,*

$$\mu(\{x : \ f(x) > 2\lambda, \ g(x) \leq \gamma\lambda\}) \leq \epsilon\mu(\{x : \ f(x) > \lambda\}). \tag{2.19}$$

Let $0 < p < \infty$. If $f \in L^p(\Omega, \mu)$ then

$$\int_\Omega (f(x))^p \, d\mu(x) \leq C \int_\Omega (g(x))^p \, d\mu(x),$$

where the finite constant C only depends on p and the way that ϵ depends on γ (to be made explicit below).

Proof of Theorem 2.3. Suppose that 2.19 holds. Then, for any λ,

$$\mu(\{x : \ f(x) > 2\lambda\}) \leq \mu(\{x : \ g(x) > \gamma\lambda\}) + \mu(\{x : \ f(x) > 2\lambda, \ g(x) \leq \gamma\lambda\})$$
$$\leq \mu(\{x : \ g(x) > \gamma\lambda\}) + \epsilon\mu(\{x : \ f(x) > \lambda\}).$$

If we multiply both sides of this inequality by $p\lambda^{p-1}$, integrate from 0 to infinity, and do two easy changes-of-variable, we get:

$$2^{-p} \int_\Omega f(x)^p \, d\mu(x) \leq \gamma^{-p} \int_\Omega g(x)^p \, d\mu(x) + \epsilon \int_\Omega f(x)^p \, d\mu(x).$$

Let $\epsilon = 2^{-p-1}$ and pick the appropriate γ. When we subtract and divide (which is okay, because $f \in L^p$), we get

$$\int_\Omega f(x)^p \, d\mu(x) \le 2^{p+1} \gamma^{-p} \int_\Omega g(x)^p \, d\mu(x),$$

which was to be proved.

The proof of Theorem 2.3 goes by so fast, it can be hard to see the idea behind it. Rewrite 2.19 for $\lambda = 2^k$. It is:

$$\mu(\{x : \; f(x) > 2^{k+1}, \, g(x) \le \gamma 2^k\}) \le \epsilon \mu(\{x : \; f(x) > 2^k\}). \qquad (2.20)$$

Imagine that f and g are having a race. For any x, let $F(x)$ be the set $\{k : \; f(x) > 2^k\}$ and let $G(x)$ be the set $\{k : \; g(x) > \gamma 2^k\}$. There are two possibilities: $F(x) \subset G(x)$, or $G(x)$ is a proper subset of $F(x)$. If the first happens, then $f(x) \le 2\gamma^{-1} g(x)$, and we can say that g is "keeping pace" with $f(x)$ (up to a factor of $2\gamma^{-1}$). If the second happens, then at some point f has clearly pulled ahead of g: there is a k such that $k \in F(x) \cap G(x)$, but $k+1 \in F(x) \setminus G(x)$: as if to say, g kept pace with f to stage k, but failed to do so at stage $k+1$. The meaning of 2.19 (or of 2.20) is that the set of x's for which this happens is *negligible*.

The proof of Theorem 2.1 will go more smoothly with the help of a useful definition, one that will follow us through later chapters.

Definition 2.2. *If N is a positive integer, we set*

$$\mathcal{D}(N) = \{I \in \mathcal{D} : \; I \subset [-2^N, 2^N), \; \ell(I) > 2^{-N}\},$$

where $\ell(I)$ is I's length.

In plain language, $\mathcal{D}(N)$ consists of the dyadic intervals that aren't too small, aren't too big, and aren't too far from the origin.

We observe that each $\mathcal{D}(N)$ is finite, $\mathcal{D}(N) \subset \mathcal{D}(N+1)$ for all N, and $\cup_N \mathcal{D}(N) = \mathcal{D}$.

We are *finally* ready to prove Theorem 2.1. (After such a long digression, the reader might want to review the definition of the dyadic square function, given by 2.8.)

Proof of Theorem 2.1. We will actually prove that, for all finite sums $f = \sum \lambda_I h_{(I)}$ and $0 < p < \infty$,

$$c_p \|f^*\|_p \le \|S(f)\|_p \le C_p \|f^*\|_p. \qquad (2.21)$$

This implies Theorem 2.1. To see this, let N be an integer, and define

$$f_N \equiv \sum_{I: \; I \in \mathcal{D}(N)} \lambda_I h_{(I)}.$$

If we prove 2.21, we will have

$$c_p \|f_N\|_p \leq \|S(f_N)\|_p \leq \|S(f)\|_p.$$

Let's set $J_N^+ = [0, 2^N)$ and $J_N^- = [-2^N, 0)$. The reader who has understood why 2.7 is true will have little trouble seeing that

$$f_N = \left(\sum_{\substack{I: \ I \subset J_N^+ \cup J_N^- \\ \ell(I) = 2^{-N}}} f_I \chi_I \right) - \left(f_{J_N^+} \chi_{J_N^+} + f_{J_N^-} \chi_{J_N^-} \right)$$

$$= (I_N) - (II_N).$$

If $f \in L^p(\mathbf{R})$ $(1 < p < \infty)$, then $(I_N) \to f$ almost everywhere, and $(II_N) \to 0$ uniformly; and so the left-hand inequality of 2.12 would follow from Fatou's Lemma.

As for the right-hand side: If $f \in L^p$ $(1 < p < \infty)$, then $|f_{J_N^+} \chi_{J_N^+} + f_{J_N^-} \chi_{J_N^-}|$ converges to 0 in L^p as $N \to \infty$. Now, for any N,

$$f_N^* \leq f^* + |f_{J_N^+} \chi_{J_N^+} + f_{J_N^-} \chi_{J_N^-}|.$$

Thus, if we prove 2.21 for finite sums, we'll have

$$\|S(f_N)\|_p \leq C_p \| f^* + |f_{J_N^+} \chi_{J_N^+} + f_{J_N^-} \chi_{J_N^-}| \|_p \tag{2.22}$$

for all N. Letting N go to infinity, we get, by Monotone Convergence,

$$\|S(f)\|_p \leq C_p \|f^*\|_p$$

for *arbitrary* $f \in L^p$, from which 2.12 follows by Theorem 2.2.

So, it's enough to show 2.21 for *finite* linear sums.

At the outset let's observe that, for all f in our test class (finite linear combinations of Haar functions), f^* and $S(f)$ are bounded and have compact support, and so belong to L^p for all p.

Proof of the first inequality in 2.21. By the good-λ inequality, it will be enough to show that, for all $\epsilon > 0$, there is a $\gamma > 0$, such that, for all $\lambda > 0$ and all f in our test class,

$$|\{x: \ f^*(x) > 2\lambda, \ S(f)(x) \leq \gamma\lambda\}| \leq \epsilon |\{x: \ f^*(x) > \lambda\}|. \tag{2.23}$$

Let $\{I_i^\lambda\}$ be the maximal dyadic intervals such that $|f_{I_i^\lambda}| > \lambda$. The intervals I_i^λ are disjoint and $\{x: \ f^*(x) > \lambda\} = \cup I_i^\lambda$. Therefore (modulo the usual business about quantifiers), it is enough to show that, for every i,

$$|\{x \in I_i^\lambda: \ f^*(x) > 2\lambda, \ S(f)(x) \leq \gamma\lambda\}| \leq \epsilon |I_i^\lambda|. \tag{2.24}$$

Fix the interval I_i^λ, and let J_i be its "dyadic double"; i.e., the unique dyadic interval for which I_i^λ is either a right or left half. Because of I_i^λ's maximality, $|f_{J_i}| \leq \lambda$. Now, if $|f_{I_i^\lambda}| > 1.1\lambda$, then $|f_{J_i} - f_{I_i^\lambda}| > .1\lambda$, and so $S(f) > .1\lambda$ on *all of* I_i^λ. In this case, if we pick $\gamma \leq .1$, the left-hand side of 2.24 is zero, and we have nothing to prove. *Henceforth we assume that $\gamma \leq .1$; and, without loss of generality, we may also assume that $|f_{I_i^\lambda}| \leq 1.1\lambda$.* Given this, our problem now reduces to showing

$$|\{x \in I_i^\lambda : (f - f_{I_i^\lambda})^*(x) > .9\lambda, \, S(f)(x) \leq \gamma\lambda\}| \leq \epsilon|I_i^\lambda|. \tag{2.25}$$

Set

$$g(x) = \begin{cases} f(x) - f_{I_i^\lambda} & \text{if } x \in I_i^\lambda; \\ 0 & \text{otherwise.} \end{cases}$$

If $I \subset I_i^\lambda$, then $\lambda_I = \langle f, h_{(I)} \rangle = \langle g, h_{(I)} \rangle$, and if $I \not\subset I_i^\lambda$, then $\langle g, h_{(I)} \rangle = 0$. (Work it out!) Therefore $S(g) \leq S(f)$ in I_i^λ and is equal to 0 outside I_i^λ. This reduces our problem to showing

$$|\{x \in I_i^\lambda : g^*(x) > .9\lambda, \, S(g)(x) \leq \gamma\lambda\}| \leq \epsilon|I_i^\lambda|. \tag{2.26}$$

Let $\{Q_j\}$ be the maximal dyadic subintervals of I_i^λ such that

$$\sum_{J: \, Q_j \subset J} \frac{|\lambda_J|^2}{|J|} > (\gamma\lambda)^2.$$

Define ('s' stands for "stopped"):

$$g_s(x) = \begin{cases} g_{Q_j} & \text{if } x \in Q_j; \\ g(x) & \text{if } x \notin \cup_j Q_j. \end{cases}$$

I claim that $S(g_s) \leq \gamma\lambda$ on all of I_i^λ, and that $(g_s)_I = g_I$ for all dyadic intervals $I \subset I_i^\lambda$ which are not *properly* contained in any Q_j. This claim, to put it mildly, requires some justification, which we reserve for the end of the proof. For now, let us take the claim for granted.

Suppose that $g^*(x) > .9\lambda$ and $S(g)(x) \leq \gamma\lambda$. Then there is an $I \subset I_i^\lambda$ such that $|g_I| > .9\lambda$. This I is either properly contained in some Q_j or it isn't. If it is contained in a Q_j, then $S(g) > \gamma\lambda$ *on all of I*, which we are assuming does NOT happen, because $S(g)(x) \leq \gamma\lambda$. Therefore: I *is NOT properly contained in any Q_j*. The second part of the claim now implies that $|(g_s)_I| > .9\lambda$. Therefore,

$$\{x \in I_i^\lambda : g^*(x) > .9\lambda, \, S(g)(x) \leq \gamma\lambda\} \subset \{x \in I_i^\lambda : g_s^*(x) > .9\lambda\},$$

with the corresponding inequality in measures.

But now we can write:

$$\int_{I_i^\lambda} (g_s^*)^2\, dx \le C \int |g_s|^2\, dx$$

$$\le C \int_{I_i^\lambda} (S(g_s))^2\, dx$$

$$\le C(\gamma\lambda)^2 |I_i^\lambda|,$$

where the last inequality follows from the first part of the claim. Now Chebyshev's inequality implies

$$|\{x \in I_i^\lambda : g_s^*(x) > .9\lambda\}| \le C(\gamma/.9)^2 |I_i^\lambda|,$$

finishing the proof.

The justification of the claim is not hard. It is an application of 2.5 and 2.4, where our intervals Q_j are the J_k's, and g_s is f_2. If $I \subset I_i^\lambda$ and I is not a subset (proper or otherwise) of any Q_j, then $\lambda_I(g_s) = \lambda_I(g)$. On the other hand, if I is a subset of a Q_j, then $\lambda_I(g_s) = 0$ (because g_s is constant across Q_j). Therefore, if $x \in Q_j$,

$$(S(g_s)(x))^2 = \sum_{\substack{J: \, Q_j \subset J \\ J \ne Q_j}} \frac{|\lambda_J|^2}{|J|} \le (\gamma\lambda)^2,$$

because of Q_j's maximality; and, if $x \notin \cup_j Q_j$, then $S(g_s)(x) \le S(g)(x) \le \gamma\lambda$. That is the first part of the claim. The second part is simply a translation of 2.4.

Proof of the second inequality in 2.21. We'll need another maximal function. Remember that I_l and I_r denote the left and right halves of a dyadic interval I. Define the "look-ahead" maximal function $L_a(f)$ by:

$$L_a(f)(x) \equiv \sup_{I:x\in I} \max(|f_{I_l}|, |f_{I_r}|).$$

It's clear that $f^*(x) \le L_a(f)(x)$ pointwise. However, a moment's thought also shows that, for all $\lambda > 0$,

$$|\{x : \, L_a(f)(x) > \lambda\}| \le 2|\{x : \, f^*(x) > \lambda\}|.$$

Therefore, in order to prove the right-hand part of 2.21, it will be enough to show the appropriate good-λ inequality between $S(f)$ and $L_a(f)$.

Let $\{I_i^\lambda\}$ be the maximal dyadic intervals such that

$$\sum_{J:I_i^\lambda \subset J} \frac{|\lambda_J|^2}{|J|} > \lambda^2.$$

Clearly, $\{x : S(f) > \lambda\} = \cup_i I_i^\lambda$. Arguing as in the proof of 2.23, it is enough to show that, for each i,

$$|\{x \in I_i^\lambda : S(f)(x) > 2\lambda, L_a(f)(x) \leq \gamma\lambda\}| \leq \epsilon|I_i^\lambda|. \tag{2.27}$$

Now, for whatever γ we decide on, if $|f_{I_i^\lambda}| > \gamma\lambda$, then the left-hand side of 2.27 will be 0, and we'll have nothing to prove. Therefore, without loss of generality, we can always assume $|f_{I_i^\lambda}| \leq \gamma\lambda$.

If

$$\sum_{J:\ I_i^\lambda \subset J} \frac{|\lambda_J|^2}{|J|} > (1.1\lambda)^2, \tag{2.28}$$

then

$$|\lambda_{I_i^\lambda}| = |\langle f, h_{(I_i^\lambda)}\rangle| > .1\lambda\sqrt{|I_i^\lambda|}. \tag{2.29}$$

Let I_l and I_r denote I_i^λ's left and right halves, respectively. Inequality 2.29 implies that

$$|f_{I_l} - f_{I_r}| > .1\lambda,$$

which in turn implies that $L_a(f) > .05\lambda$ on all of I_i^λ. Therefore, if $\gamma \leq .05$ (which we henceforth assume) and 2.28 holds, then the left-hand side of 2.27 is 0 and, again, we have nothing to prove. So, in perfect analogy with the proof of 2.23, we may assume that 2.28 does *not* hold.

Let's now define

$$f_{loc} \equiv \sum_{I \subset I_i^\lambda} \lambda_I h_{(I)},$$

which we think of as the "localized" version of f. Since $|f_{I_i^\lambda}| \leq \gamma\lambda$, we have that

$$\{x \in I_i^\lambda : L_a(f)(x) \leq \gamma\lambda\} \subset \{x \in I_i^\lambda : L_a(f_{loc})(x) \leq 2\gamma\lambda\}.$$

Also, because 2.28 does *not* hold,

$$\{x \in I_i^\lambda : S(f)(x) > 2\lambda\} \subset \{x \in I_i^\lambda : S(f_{loc})(x) > .9\lambda\}.$$

Putting these two inclusions together, we see that it is enough to show that

$$|\{x \in I_i^\lambda : S(f_{loc})(x) > .9\lambda, L_a(f_{loc})(x) \leq 2\gamma\lambda\}| \leq \epsilon|I_i^\lambda|,$$

and it is this inequality which we will now prove.

Let $\{J_k\}$ be the maximal dyadic subintervals of I_i^λ having the property that either the right or left half of J_k—call it J_k^*—satisfies $|(f_{loc})_{J_k^*}| > 2\gamma\lambda$. Set (once again, 's' is for "stopped"):

$$f_s(x) = \begin{cases} (f_{loc})_{J_k} & \text{if } x \in J_k; \\ f_{loc}(x) & \text{if } x \notin \cup_k J_k. \end{cases}$$

Because of J_k's maximality, $|f_s(x)| \leq 2\gamma$ almost everywhere.

Suppose that $x \in I_i^\lambda$ is a point such that $S(f_{loc})(x) > .9\lambda$ and $L_a(f_{loc})(x) \leq 2\gamma\lambda$. Then $x \notin \cup_k J_k$. We claim that, if $x \notin \cup_k J_k$, then $S(f_s)(x) = S(f_{loc})(x)$. The proof of this is simply the fact that

$$\langle f_s, h_{(I)} \rangle = \langle f_{loc}, h_{(I)} \rangle \tag{2.30}$$

if I is not contained in any J_k—which is just equation 2.5 showing up again.

Therefore,

$$\{x \in I_i^\lambda : S(f_{loc})(x) > .9\lambda, L_a(f_{loc})(x) \leq 2\gamma\lambda\} \subset \{x \in I_i^\lambda : S(f_s)(x) > .9\lambda\}.$$

However,

$$\int_{I_i^\lambda} (S(f_s))^2 \, dx = \int_{I_i^\lambda} |f_s(x)|^2 \, dx$$
$$\leq (2\gamma)^2 |I_i^\lambda|,$$

and now our result follows (again!) by Chebyshev's inequality. This finishes the proof of 2.21, and thus of Theorem 2.1.

If we look closely at the proof of Theorem 2.1, we can see that we have obtained something slightly stronger. Recall that we proved 2.21 for all p, $0 < p < \infty$, for finite linear sums of Haar functions. The only role this finiteness hypothesis played was to ensure that $S(f)$ and f^* belonged to L^p. Therefore we have actually shown that, if $0 < p < \infty$ and $f^* \in L^p$, then

$$\|f^*\|_p \leq c_p \|S(f)\|_p; \tag{2.31}$$

and, if $S(f) \in L^p$, then

$$\|S(f)\|_p \leq C_p \|f^*\|_p. \tag{2.32}$$

It would be nice to obtain the statements $f^* \in L^p$ (respectively $S(f) \in L^p$) as conclusions of 2.31 (respectively 2.32) rather than having to assume them as hypotheses.

Unfortunately, *some* extra hypothesis is needed in 2.31: if f is identically 1, its square function is identically 0. But a case like this, it turns out, is essentially the worst that can happen. That is, if $|f_{J_N^+}|$ and $|f_{J_N^-}|$ both go to 0 as $N \to \infty$, then 2.31 holds. Because, if these averages go to 0, $f^*(x)$ will equal the limit of $f_N^*(x)$ for every x, implying, by Fatou's Lemma,

$$\int (f^*(x))^p \, dx = \int (\lim_N f_N^*(x))^p \, dx$$
$$\leq C \lim_N \int (S(f_N))^p \, dx$$
$$= C \int (S(f))^p \, dx.$$

However, we do not need to assume anything about $S(f)$ in 2.32. This is because, if $0 < p < \infty$, then, for any positive N,

$$\|S(f_N)\|_p \leq C_p \|f_N^*\|_p.$$

But $f_N^* \leq 2f^*$ everywhere; while $\|S(f_N)\|_p \nearrow \|S(f)\|_p$ because of Monotone Convergence. Therefore, if $f^* \in L^p$, $S(f) \in L^p$.

A moment's (or perhaps several moments') thought shows that weighted forms of 2.21's component inequalities,

$$\int (f^*(x))^p \, v \, dx \leq c(p, v) \int (S(f))^p \, v \, dx \tag{2.33}$$

and

$$\int (S(f))^p \, v \, dx \leq C(p, v) \int (f^*(x))^p \, v \, dx \tag{2.34}$$

should follow for any weight v that is "regular" enough. But, before we say what "regular" means here, let us first explain what we mean by a weight.

Definition 2.3. *A weight v is a non-negative, locally integrable function.*

If v is a weight and E is a measurable set, we use $v(E)$ to denote $\int_E v \, dx$. With those preliminaries out of the way, let's now introduce two classes of weights.

Definition 2.4. *A weight v is said to belong to A_∞^d (pronounced, "dyadic A-infinity" or "A-infinity dyadic") if, for every $\epsilon > 0$, there is a $\delta > 0$ such that, for all dyadic intervals I and measurable $E \subset I$, $|E|/|I| < \delta$ implies $v(E) \leq \epsilon v(I)$.*

Remark. If the preceding implications hold for *all* bounded intervals I, v is said to belong to A_∞.

Definition 2.5. *A weight v is said to be* dyadic doubling *if there are positive constants c and C such that, for all dyadic intervals I,*

$$cv(I_l) \leq v(I_r) \leq Cv(I_l).$$

Remark. If

$$cv(I) \leq v(I') \leq Cv(I)$$

for any two congruent intervals that touch, v is said to be *doubling*. This is equivalent to saying that $v(2I) \leq Cv(I)$ for some C, for all intervals I, where $2I$ means the concentric double of I. If v is defined on $\mathbf{R}^d$, v is said to be doubling if $v(2Q) \leq Cv(Q)$ for all cubes $Q \subset \mathbf{R}^d$, and it is said to be dyadic doubling if, for every $Q \in \mathcal{D}_d$, $v(Q)$ is bounded by a constant times $v(Q')$ for any dyadic $Q' \subset Q$ such that $\ell(Q') = (1/2)\ell(Q)$. The reader might want to check that this reduces to Definition 2.5 when $d = 1$.

What do these definitions mean? They are two different ways of saying that a weight v is "almost constant." Notice that they apply to constant weights and to weights v for which $v(x)$ and $1/v(x)$ are both bounded functions. They also apply to weights of the form $|x|^r$, where r is any positive real number. With a little work the reader can see that they even apply to certain weights $|x|^r$ for negative r.

Essentially, the property of belonging to A_∞^d means that, considered as a measure, v is absolutely continuous, uniformly, with respect to changes in scale. Another way to say this is that, if $v \in A_\infty^d$, then, on any $I \in \mathcal{D}$, v does not put very much of its mass, relative to I, on very small subsets of I.

We have the following characterization of A_∞^d, which we will prove in the next chapter (Theorem 3.3):

Theorem 2.4. *Let v be a weight. Then $v \in A_\infty^d$ if and only if there is a positive constant A such that, for all dyadic intervals I,*

$$\int_I M_d(v\chi_I)\,dx \le Av(I). \tag{2.35}$$

This characterization yields a cornucopia of A_∞^d weights. Here is how. Suppose that a weight w belongs to L^p, with $1 < p \le \infty$. We know that there is a constant C_p such that $\|M_d(f)\|_p \le C_p\|f\|_p$ for all $f \in L^p$. I claim that there is a weight $\tilde{w}$ such that

$$w(x) \le C_{1,p}\tilde{w}(x) \text{ everywhere} \tag{2.36}$$

$$\|\tilde{w}\|_p \le C_{2,p}\|w\|_p \tag{2.37}$$

$$M_d(\tilde{w})(x) \le C_{3,p}\tilde{w}(x) \text{ a.e.} \tag{2.38}$$

The first two inequalities say that w and $\tilde{w}$ are roughly the same size. The third inequality implies that $\tilde{w}$ satisfies 2.35.

There are at least two ways to construct such a weight $\tilde{w}$. The first method, due to José Luis Rubio de Francia, is theoretically elegant, but a little hard to compute. Simply set:

$$\tilde{w}(x) \equiv w(x) + (2C_p)^{-1}M_d(w)(x) + (2C_p)^{-2}M_d^2(w)(x) + \cdots, \tag{2.39}$$

where M_d^k denotes a k-fold application of the operator M_d. This yields a $\tilde{w}$ for which $C_{1,p} \le 1$, $C_{2,p} \le 2$, and $C_{3,p} \le 2C_p$: 2.36 is trivial; 2.37 follows from the operator bound on M_d; and, applying M_d to both sides of 2.39 yields

$$M_d(\tilde{w})(x) \le M_d(w)(x) + (2C_p)^{-1}M_d^2(w)(x) + (2C_p)^{-2}M_d^3(w)(x) + \cdots$$
$$\le 2C_p\tilde{w}(x),$$

which is 2.38.

The second method is cruder but, in many cases, more computable. It makes use of the following lemma, which we will also prove in the next chapter.

Lemma 2.1. *If v and $M_d(v)$ are weights, and $0 < \beta < 1$, then*

$$M_d((M_d(v))^\beta) \leq C_\beta (M_d(v))^\beta$$

almost everywhere.

How does this give us a weight like $\tilde{w}$? Suppose $w \in L^p$, $p > 1$. Let $1 < r < p$, and set $v = w^r$ and $\beta = 1/r < 1$. Then $M_d(v)$ is a weight, because it belongs to $L^{p/r}$. Indeed:

$$\|M_d(v)\|_{p/r}^r \leq C_{p,r} \|v\|_{p/r}^r$$

$$= C_{p,r} \int (w(x))^p \, dx.$$

We can then take $\tilde{w}$ to be $(M_d(v))^\beta$. The advantage of this construction is that it yields $\tilde{w}$ through the application of only *one* maximal function. One of its disadvantages is that this $\tilde{w}$ can easily be a lot bigger than the one given by 2.39.

Notice that both of these constructions, by making use of maximal functions, tend to "homogenize" the weight. This is consistent with our observation that A_∞^d weights are "almost constant."

We can see this homogenization in action if we apply the second construction to $w = \chi_{[0,1)} + \chi_{[n-1,n)}$, where $n >> 1$. Put $v = w^r$, where $r > 1$ is fixed. Observe that $M_d(v) \equiv 0$ on $(-\infty, 0)$. The function $(M_d(v))^\beta$ has a value of 1 on $[0, 1)$, then decreases, to bottom out at around $n^{-\beta}$, near $x = n/2$. It then rises, to equal 1 again on $[n, n+1)$, after which it decreases toward 0, being comparable to $|x - n|^{-\beta}$ when x is large.

The result of the construction has been, in effect, to drape a tent over the graph of w. A good exercise for the reader is to show that these sorts of constructions ("finding L^p, A_∞^d majorants") are impossible in L^1. In other words, the reader should show that if v is an A_∞^d weight, and v is not identically 0, then $v \notin L^1(\mathbf{R})$.

The property of being dyadic doubling means that, if v assigns positive mass to any dyadic interval I, it assigns roughly equal positive masses to I's right and left halves. One consequence of this property is that if v is dyadic doubling and $v(I) > 0$ for any dyadic $I \subset [0, \infty)$, then $v(I) > 0$ for *every* dyadic $I \subset [0, \infty)$, with the analogous fact holding for subintervals of $(-\infty, 0)$.

It is easy to see that $v \in A_\infty^d$ does not imply that v is dyadic doubling; and the reader is invited to build a counterexample showing this. If we require that the $\epsilon - \delta$ implication should hold for *all* intervals I, then we do get dyadic doubling; and, indeed, doubling, i.e., the existence of a constant C such that, for all intervals I, $v(2I) \leq Cv(I)$, where $2I$ is I's *concentric* double. This is not especially hard to prove, and we have made this an exercise in the next chapter. It is harder to show that the dyadic doubling property does not imply $v \in A_\infty^d$; exercise 14.1 outlines a proof of this fact.

The relevance of these classes to weighted problems is:

Theorem 2.5. *If $v \in A_\infty^d$, then 2.33 holds for all finite linear sums of Haar functions.*

Theorem 2.6. *If $v \in A_\infty^d$ and is dyadic doubling, then 2.34 holds for all finite linear sums of Haar functions.*

If the reader goes over the proof of 2.21, he will see that the v-weighted version of the good-λ inequality, namely

$$v\left(\{x \in I_i^\lambda : \ f^*(x) > 2\lambda, \ S(f)(x) \le \gamma\lambda\}\right) \le \epsilon v(I_i^\lambda)$$

(and that's all that's needed for Theorem 2.5), only requires $v \in A_\infty^d$. Dyadic doubling yields

$$v(\{L_a(f) > \lambda\}) \le \tilde{C} v(\{f^* > \lambda\}),$$

which is the extra element needed in the proof of Theorem 2.6.

What can we say if $v \notin A_\infty^d$? In that case it is impossible for the conclusion of Theorem 2.5 to hold; we shall see a proof of this later. As to whether Theorem 2.6 holds, the author doesn't know. In any event, the general inequalities we *can* prove when $v \notin A_\infty^d$ have the form

$$\int (f^*(x))^p \, v \, dx \le C \int (S(f))^p \, w \, dx \tag{2.40}$$

and

$$\int (S(f))^p \, v \, dx \le C \int |f(x)|^p \, w \, dx \tag{2.41}$$

for pairs of weights (v, w). A complete "solution" to the problem presented by 2.40 (respectively 2.41) would be a set of *testable* necessary and sufficient conditions on pairs (v, w) for inequality 2.40 (respectively 2.41) to hold for all finite linear combinations $f = \sum \lambda_I h_{(I)}$. Such an achievement is probably too much to hope for, at least in this lifetime. We will content ourselves with obtaining strong *sufficient* conditions for these two inequalities.

By setting $f = h_{(I)}$, for I arbitrary, we see that a necessary condition for either 2.40 or 2.41 to hold is that $v \le Cw$ almost everywhere. It is therefore natural to look for sufficient conditions that automatically imply $v \le Cw$. We have already seen *two* such sufficient conditions, though not their proofs (which will come in the next chapter). If R is any number larger than 1, we define the *Rubio de Francia maximal function of v*, with parameter R, by the following equation:

$$\mathbf{M}_R(v) \equiv \sum_0^\infty \frac{M_d^k(v)}{R^k},$$

where $M_d^0(v)$ is just v. Likewise, if $r > 1$ and v is a weight, let us define

$$M_{r,d}(v) \equiv (M_d(v^r))^{1/r}.$$

Both of these functions are pointwise $\geq v$. We know that the first one satisfies the inequality $M_d(w) \leq Cw$, and we have asserted (but not yet proved) that the second one does too. We have also asserted (but not yet proved) that, if w is any weight satisfying

$$M_d(w) \leq Cw, \tag{2.42}$$

then $w \in A_\infty^d$. Therefore, if we can prove 2.42 for $M_{r,d}(v)$, and the implication "2.42 $\Rightarrow$ the weight w belongs to A_∞^d," we will have:

Theorem 2.7. *a) If v and $\mathbf{M}_R(v)$ are weights, then for all $0 < p < \infty$ and all $f = \sum_I \lambda_I h_{(I)}$, finite linear sums of Haar functions,*

$$\int (f^*(x))^p \, v \, dx \leq C_{p,R} \int (S(f))^p \, \mathbf{M}_R(v) \, dx \tag{2.43}$$

and

$$\int (S(f))^p \, v \, dx \leq C_{p,R} \int (L_a(f)(x))^p \, \mathbf{M}_R(v) \, dx. \tag{2.44}$$

b) If v and $M_{r,d}(v)$ are weights, then for all $0 < p < \infty$ and all $f = \sum_I \lambda_I h_{(I)}$, finite linear sums of Haar functions,

$$\int (f^*(x))^p \, v \, dx \leq C_{p,r} \int (S(f))^p \, M_{r,d}(v) \, dx \tag{2.45}$$

and

$$\int (S(f))^p \, v \, dx \leq C_{p,r} \int (L_a(f)(x))^p \, M_{r,d}(v) \, dx. \tag{2.46}$$

The proofs of these inequalities are fast and easy. For 2.43, we write,

$$\int (f^*(x))^p \, v \, dx \leq \int (f^*(x))^p \, \mathbf{M}_R(v) \, dx$$

$$\leq C_{p,R} \int (S(f))^p \, \mathbf{M}_R(v) \, dx,$$

where the first inequality comes from the fact that $v \leq \mathbf{M}_R(v)$ and the second is due to the A_∞^d property of $\mathbf{M}_R(v)$. The proofs of the others are practically identical.

Corollary 2.1. *a) If v and w are two weights and $\mathbf{M}_R(v) \leq w$ almost everywhere for some $R > 1$, then*

$$\int (f^*(x))^p \, v \, dx \leq C_{p,R} \int (S(f))^p \, w \, dx$$

and

$$\int (S(f))^p \, v \, dx \leq C_{p,R} \int (L_a(f)(x))^p \, w \, dx$$

hold for all $0 < p < \infty$ and all finite linear sums $f = \sum_I \lambda_I h_{(I)}$. b) If v and w are two weights and $M_{r,d}(v) \leq w$ almost everywhere for some $r > 1$, then

$$\int (f^*(x))^p \, v \, dx \le C_{p,r} \int (S(f))^p \, w \, dx$$

and

$$\int (S(f))^p \, v \, dx \le C_{p,r} \int (L_a(f)(x))^p \, w \, dx$$

hold for all $0 < p < \infty$ *and all finite linear sums* $f = \sum_I \lambda_I h_{(I)}$.

In essence, Theorem 2.7 and Corollary 2.1 say that (v, w) is a good pair for our weighted problems if w is bigger than or equal to "enough" (infinitely many, but rapidly damped down) iterates of M_d applied to v, or a "bumped up" version of $M_d(v)$ (bumped up to $(M_d(v^r))^{1/r}$). It is reasonable to ask: Can we get away with just v? If not, is plain old unbumped M_d, applied to an arbitrary weight v, enough?

The answer to the first reasonable question is "definitely not." For $k = 0, 1, 2, \ldots$, let $I_k = [0, 2^k)$, and set $\lambda_k = 2^{k/2}/(k+1)$. For $N \ge 0$, define

$$g_N \equiv \sum_0^N \lambda_k h_{(I_k)}.$$

Then $\|S(g_N)\|_\infty \le (\sum_0^\infty (k+1)^{-2})^{1/2}$, independent of N, but $g_N(x) = \sum_0^N (k+1)^{-1} \nearrow \infty$ for all $x \in [0, 1)$. Therefore

$$\int |g_N|^p \, v \, dx \le C \int S^p(g_N) \, v \, dx$$

cannot hold, for any p, with a constant C independent of v and N: just set $v = \chi_{[0,1)}$. Similarly,

$$\int (S(f))^p \, v \, dx \le C_p \int (L_a(f))^p \, v \, dx$$

cannot hold with a constant independent of f and v. To see this, again let $v = \chi_{[0,1)}$. For $k \ge 0$, define $\tilde{\lambda}_k \equiv 2^{k/2}/\sqrt{k+1}$. Using the same I_k's as before, define, for $N \ge 0$,

$$\tilde{g}_N \equiv \sum_0^N (-1)^k \tilde{\lambda}_k h_{(I_k)}.$$

The alternating factor $(-1)^k$ ensures that $\|\tilde{g}_N\|_\infty$ (hence $\|L_a(\tilde{g}_N)\|_\infty$) has a bound independent of N. Hence

$$\int (L_a(\tilde{g}_N)(x))^p \, v \, dx \le C_p.$$

But

$$\int (S(\tilde{g}_N))^p \, v \, dx = \left(\sum_0^N (k+1)^{-1} \right)^{p/2} \to \infty.$$

The answer to the second question—about whether $M_d(v)$ is ever big enough—is, "sometimes yes and sometimes no—but 'no' is the safer bet." We will see this in the next chapter. At the same time we will learn some pretty good sufficient conditions on pairs of weights (v, w) which ensure that 2.40 or 2.41 holds for all f in some large test class, with constants independent of f.

The search for these conditions will be somewhat circuitous. It will begin with an examination of what is probably the least forceful link in the proofs of Theorem 2.5 and Theorem 2.6: namely, the weak-type $L^2 - L^2$ bounds between f and $S(f)$. At present we know that, if f (respectively, $S(f)$) is bounded, then, on any bounded interval, the measure of the set where $S(f)$ (respectively, $|f|$) is bigger than λ decays at least as fast as λ^{-2}.

We will see that these measures actually decay a lot faster.

Mini-Appendix: Interpolation

We will now give the promised proof of (a special case of) the Marcinkiewicz Interpolation Theorem (Theorem 2.8).

Definition 2.6. *An operator T is called subadditive if $|T(f + g)| \leq |Tf| + |Tg|$ pointwise for all f and g such that Tf, Tg, and $T(f + g)$ make sense. The operator T is called homogeneous if, for all scalars α and all f in T's domain, $|T(\alpha f)| = |\alpha||Tf|$. If T is subadditive and homogeneous, it is called sublinear.*

Definition 2.7. *Let $(X, \mathcal{M}, \mu)$ and $(Y, \mathcal{N}, \nu)$ be two measure spaces, $0 < p < \infty$, and suppose that T is a sublinear operator mapping from $L^p(X, \mathcal{M}, \mu)$ into the space of Y-measurable functions. T is said to be of weak type (p, p) if there is a constant C such that, for all f and for all $\lambda > 0$,*

$$\nu\left(\{y \in Y : |Tf(y)| > \lambda\}\right) \leq C\lambda^{-p} \int_X |f(x)|^p \, d\mu(x).$$

T is said to be of weak type (∞, ∞) if there is a constant C such that $\|Tf\|_{L^\infty(Y)} \leq C\|f\|_{L^\infty(X)}$ for all bounded f.

We note (and the reader should show) that if T maps boundedly from $L^p(X)$ into $L^p(Y)$, then T is automatically weak type (p, p). The converse is false. We proved that M_d is weak type $(1, 1)$, but the example of $f = \chi_{[0,1)}$ shows that it is not bounded on L^1.

Theorem 2.8. *Suppose that $(X, \mathcal{M}, \mu)$ and $(Y, \mathcal{N}, \nu)$ are two σ-finite measure spaces. Let T be a sublinear operator as defined above and let $0 < p_1 < p < p_2 \leq \infty$. If T is of weak type (p_1, p_1) and (p_2, p_2), then it maps boundedly from $L^p(X)$ into $L^p(Y)$.*

Proof. Let $f \in L^p$, fix $\lambda > 0$, and write $f = f_1 + f_2$, where

$$f_1(x) = \begin{cases} f(x) & \text{if } |f(x)| > \lambda; \\ 0 & \text{otherwise.} \end{cases}$$

Notice (prove!) that $f_1 \in L^{p_1}$ and $f_2 \in L^{p_2}$. Let's first assume that $p_2 < \infty$. Because T is sublinear,

$$\nu\left(\{y \in Y : |Tf(y)| > \lambda\}\right)$$
$$\leq \nu\left(\{y \in Y : |Tf_1(y)| > \lambda/2\}\right) + \nu\left(\{y \in Y : |Tf_2(y)| > \lambda/2\}\right).$$

The first term following the '$\leq$' sign is no bigger than

$$\frac{C}{\lambda^{p_1}} \int_{|f|>\lambda} |f|^{p_1} \, d\mu(x), \tag{2.47}$$

while the second term is no bigger than

$$\frac{C}{\lambda^{p_2}} \int_{|f|\leq\lambda} |f|^{p_2} \, d\mu(x). \tag{2.48}$$

If we multiply 2.47 by $p\lambda^{p-1}$ and integrate from 0 to ∞, we get a constant times

$$\int_0^\infty \lambda^{p-1-p_1} \left(\int_{|f|>\lambda} |f| \, d\mu(x) \right) d\lambda = \int_X |f|^{p_1} \left(\int_0^{|f|} \lambda^{p-1-p_1} \, d\lambda \right) d\mu(x)$$
$$= (p-p_1)^{-1} \int_X |f|^{p_1} |f|^{p-p_1} \, d\mu(x)$$
$$= (p-p_1)^{-1} \int_X |f|^p \, d\mu(x),$$

and that's what we want. Notice how crucial it was for $p - 1 - p_1$ to be larger than -1.

If we perform similar manipulations on 2.48 we get a constant times

$$\int_X |f|^{p_2} \left(\int_{|f|}^\infty \lambda^{p-1-p_2} \, d\lambda \right) d\mu(x);$$

which (because $p - 1 - p_2$ is *less* than -1) is equal to

$$(p_2 - p)^{-1} \int_X |f|^{p_2} |f|^{p-p_2} \, d\mu(x) = (p_2 - p)^{-1} \int_X |f|^p \, d\mu(x),$$

and that's also okay.

If $p_2 = \infty$, we split f at $\epsilon\lambda$ (instead of λ), where ϵ is positive, but chosen so small that $\{y \in Y : |Tf_2(y)| > \lambda/2\}$ has ν-measure equal to 0. The theorem is proved.

Exercises

2.1. Prove equation 2.14. (Hint: Follow this trail: characteristic functions $\rightarrow$ simple functions $\rightarrow$ monotone limits of simple functions.)

2.2. Prove 2.15 and 2.16 without the assumption that $|f|_I = |I| = 1$.

2.3. The Hardy-Littlewood maximal function of f, a locally integrable function, is defined by

$$M(f)(x) \equiv \sup_{I:x\in I} \frac{1}{|I|} \int_I |f(t)|\, dt,$$

where the supremum is over all bounded intervals containing x. It is trivial that $M_d(f) \leq M(f)$ pointwise. Show that there are positive constants c_1 and c_2 such that, for all f and all $\lambda > 0$,

$$|\{x:\ M(f)(x) > \lambda\}| \leq c_1 |\{x:\ M_d(f)(x) > c_2\lambda\}|,$$

with the consequence that, for all $1 < p \leq \infty$,

$$\|M(f)\|_p \leq C_p \|f\|_p.$$

2.4. We can generalize M_d and M to d dimensions as follows. If $f : \mathbf{R}^d \mapsto \mathbf{R}$ is measurable, then we set

$$M_d(f)(x) \equiv \sup_{\substack{Q\in\mathcal{D}_d \\ x\in Q}} \frac{1}{|Q|} \int_Q |f|\, dt$$

and

$$M(f)(x) \equiv \sup_{\substack{Q \text{ a cube} \\ x\in Q}} \frac{1}{|Q|} \int_Q |f|\, ds.$$

Trivially, $M_d(f) \leq M(f)$ pointwise. Show that, for all $1 < p \leq \infty$,

$$\|M(f)\|_p \leq C_{p,d} \|f\|_p.$$

2.5. State and prove d-dimensional versions of 2.15 and 2.16, in which M_d is replaced by the operator M, and the interval I is replaced by an arbitrary d-dimensional cube Q.

2.6. If $t \in \mathbf{R}^d$ and $y > 0$, the ball $B(t;y)$ is the set of $x \in \mathbf{R}^d$ such that $|x - t| < y$. The classical Hardy-Littlewood maximal operator, M_C, is defined by

$$M_C(f)(x) \equiv \sup_{\substack{(t,y)\in\mathbf{R}^{d+1}_+ \\ x\in B(t;y)}} \frac{1}{|B(t;y)|} \int_{B(t;y)} |f|\, ds.$$

Show that there are positive constants c_1 and c_2, depending only on d, such that, for all measurable f and all $x \in \mathbf{R}^d$,

$$c_1 M_C(f)(x) \leq M(f)(x) \leq c_2 M_C(f)(x),$$

and that therefore $\|M_C(f)\|_p \leq C_{p,d} \|f\|_p$ for all $1 < p \leq \infty$.

2.7. Let $h : [0, \infty) \mapsto [0, \infty]$ be non-increasing. (Notice that we allow h to be infinite.) Define $\Phi : \mathbf{R}^d \mapsto [0, \infty]$ by $\Phi(x) \equiv h(|x|)$, and suppose that $\int_{\mathbf{R}^d} \Phi(x) \, dx = 1$. Show that, if $g : \mathbf{R}^d \mapsto \mathbf{R}$ is non-negative and measurable, then, for all $x \in \mathbf{R}^d$ and all $y > 0$,

$$g * \Phi_y(x) \leq M_C(g)(x). \tag{2.49}$$

(Hint: It's enough to prove this when $x = 0$ and $y = 1$. You might have an easier time seeing what's going on if you try to prove something stronger; namely, that

$$g * \Phi(0) \leq \sup_{y > 0} \frac{1}{|B(0; y)|} \int_{B(0; y)} g(s) \, ds.)$$

If we replace 0 by an arbitrary point x, the supremum on the right is called the *centered Hardy-Littlewood maximal function of g*, evaluated at the point x. An important special case of 2.49 is when $\Phi(x) = c_{\beta,d}(1 + |x|)^{-d-\beta}$ for some $\beta > 0$. For this particular Φ we actually have (and the reader should show it):

$$\sup_{(t,y): \, |x-t| < y} g * \Phi_y(t) \leq C(\beta, d) M_C(g)(x).$$

In other words, the inequality isn't spoiled even if we "jiggle" the kernel Φ a bit. We use this fact in the proof of Theorem 6.1.

2.8. Suppose that v is a weight defined on $\mathbf{R}^d$ and f is a non-negative measurable function. For every cube $Q \subset \mathbf{R}^d$, set

$$f_{v,Q} \equiv \begin{cases} \frac{1}{v(Q)} \int_Q f(t) \, v(t) \, dt & \text{if } v(Q) > 0; \\ 0 & \text{if } v(Q) = 0. \end{cases}$$

Define

$$M_{v,d}(f)(x) \equiv \sup_{Q: \, x \in Q \in \mathcal{D}_d} f_{v,Q}.$$

This is the *v-weighted dyadic maximal function* of f. We can similarly define

$$M_v(f)(x) \equiv \sup_{\substack{Q: \, x \in Q \\ Q \text{ a cube}}} f_{v,Q},$$

the *v-weighted maximal function* of f. a) Show that

$$\|M_{v,d}(f)\|_{L^p(\mathbf{R}^d, v)} \leq C_p \|f\|_{L^p(\mathbf{R}^d, v)}$$

for all $1 < p \leq \infty$, where the constant C_p only depends on p and not on d or v. b) Suppose that v has the doubling property; i.e., that $v(2Q) \leq Cv(Q)$ for all cubes Q. Show that

$$\|M_v(f)\|_{L^p(\mathbf{R}^d,v)} \leq C\|M_{v,d}(f)\|_{L^p(\mathbf{R}^d,v)},$$

for all $1 < p \leq \infty$, where the constant C now depends on p, d, and v.

2.9. Recall our definition of f_N:

$$f_N = \sum_{I \in \mathcal{D}(N)} \lambda_I(f)h_{(I)}.$$

Show that, if $f \in L^p$ ($1 < p < \infty$), then $f_N \to f$ in the L^p norm. Show that this implication fails for $p = 1$ and $p = \infty$.

2.10. Now define

$$m(f)_N = \sum_{\substack{I \in \mathcal{D} \\ \ell(I)=2^{-N}}} f_I \chi_I;$$

i.e., this is just f replaced by its averages over dyadic intervals of length 2^{-N}. Show that, if $f \in L^p$ ($1 \leq p < \infty$), then $m(f)_N \to f$ in the L^p norm: notice that this time we're including $p = 1$. Show that convergence can still fail for $p = \infty$. (Hint: Almost-everywhere convergence comes from the Lebesgue differentiation theorem. For the norm convergence, first show that it works for continuous functions with compact support.)

2.11. Let $\mathcal{F}_1 \subset \mathcal{F}_2 \subset \mathcal{F}_3 \cdots$ be an increasing sequence of finite subsets of $\mathcal{D}$ such that $\cup_k \mathcal{F}_k = \mathcal{D}$. Elementary functional analysis implies that, if $f \in L^2$, the sequence of functions

$$f_{\mathcal{F}_k} \equiv \sum_{I \in \mathcal{F}_k} \lambda_I(f)h_{(I)} \qquad (2.50)$$

converges to f in the L^2 norm. Using Littlewood-Paley theory, we can show more: If $f \in L^p$ ($1 < p < \infty$), the sequence defined by 2.50 converges to f in the L^p norm. One argument goes like this (the missing steps are left to the reader.) We know that $\|f\|_p \sim \|S(f)\|_p$. Since $f \in L^p$, $S(f - f_{\mathcal{F}_k}) \leq S(f) < \infty$ and $S(f - f_{\mathcal{F}_k}) \to 0$ almost everywhere. Therefore, by Dominated Convergence, $S(f - f_{\mathcal{F}_k}) \to 0$ in L^p, implying the result. We could also go another way: the sequence of operators defined by $f \mapsto f_{\mathcal{F}_k}$ is uniformly bounded on L^p ($1 < p < \infty$), and the sequence converges to the identity on finite linear sums of Haar functions, which are dense in L^p.

2.12. Show that, if $v \in A_\infty^d \cap L^1$, then $v \equiv 0$.

Notes

The Haar functions first occur in [27]. The interpolation theorem (Theorem 2.8) is first proved in [40]. The Hardy-Littlewood Maximal Theorem (a general term referring to the results expressed in Theorem 2.2, exercise 2.3, exercise 2.4, exercise 2.5, and exercise 2.6) was first proved in its non-dyadic, one-dimensional form in [28]; it has since been generalized in a large but finite number of ways. The Rubio de Francia maximal function, which has many variations, first appeared (in a non-dyadic form) in [49]; see also [23]. Good-λ inequalities first appeared in the work of Burkholder and Gundy [1] [2]. This work was followed by the seminal papers [3] and [19], which established the surprising L^p equivalence between certain maximal and square functions. This equivalence was sharpened significantly in [20] and [43].

Good-λ inequalities have fallen out of fashion lately, because they are not well-suited to two-weight problems. Nevertheless, the author believes they are something every analyst should know about. One must walk before one can run, and, when even the one-weight problem looks intractable, good-λ inequalities can be just the tool one needs to pry things open.

3

Exponential Square

A function f is said to be *uniformly locally exponentially square integrable* if there exist positive constants α and β such that, for all finite intervals I,

$$\frac{1}{|I|} \int_I \exp(\alpha|f - f_I|^2)\, dx \le \beta. \tag{3.1}$$

We express this in symbols by $f \in Exp(L^2_{loc})$.

By Chebyshev's inequality, $f \in Exp(L^2_{loc})$ implies that, for all intervals I and $\lambda > 0$,

$$|\{x \in I : |f - f_I| > \lambda\}| \le \beta \exp(-\alpha\lambda^2)|I|. \tag{3.2}$$

Conversely, by integrating distribution functions, the existence of positive numbers α and β such that 3.2 holds for all $\lambda > 0$ and intervals I implies that $f \in Exp(L^2_{loc})$ (with different α and β, of course).

The good-λ inequalities used in the proof of Theorem 2.1 turned on the facts that, if f is bounded on I, then $S(f)$ is in $L^2(I)$, in a controlled way; and, if $S(f)$ is bounded on I, then f is in $L^2(I)$, in a controlled way.

But much more is true.

Theorem 3.1. *There exist positive constants α and β such that, if $\|f\|_\infty \le 1$, then for all dyadic intervals I and positive numbers λ,*

$$|\{x \in I : S(f - f_I) > \lambda\}| \le \alpha|I| \exp(-\beta\lambda^2).$$

Theorem 3.2. *There exist positive constants α and β such that, if $\|S(f)\|_\infty \le 1$, then, for all dyadic intervals I and positive numbers λ,*

$$|\{x \in I : |f - f_I| > \lambda\}| \le \alpha|I| \exp(-\beta\lambda^2).$$

Essentially, we are saying that $f \in L^\infty$ implies $S(f) \in Exp(L^2_{loc})$ with respect to *dyadic* intervals I, and vice versa.

We shall prove these theorems momentarily. Before doing so, let us note that they imply the following improved good-λ inequalities:

Corollary 3.1. *There exist positive constants C_1 and C_2 such that, for all finite linear sums $f = \sum \lambda_I h_{(I)}$ and all $\lambda > 0$,*

$$|\{x : \ S(f)(x) > 2\lambda, \ L_a(f)(x) \le \gamma\lambda\}| \le C_1 \exp(-C_2/\gamma^2)|\{x : \ S(f)(x) > \lambda\}|.$$

Corollary 3.2. *There exist positive constants C_1 and C_2 such that, for all finite linear sums $f = \sum \lambda_I h_{(I)}$ and all $\lambda > 0$,*

$$|\{x : \ f^*(x) > 2\lambda, \ S(f)(x) \le \gamma\lambda\}| \le C_1 \exp(-C_2/\gamma^2)|\{x : \ f^*(x) > \lambda\}|.$$

Remark. These corollaries, of course, follow from the analogous inequalities on the maximal dyadic intervals I_i^λ that make up $\{x : \ S(f)(x) > \lambda\}$ (respectively, $\{x : \ f^*(x) > \lambda\}$).

The proof of Theorem 3.1 is pretty short, and it follows from a simple trick. The proof of Theorem 3.2, while also fairly short, uses a cleverer trick.

Proof of Theorem 3.1. We claim that there is a constant C such that, for any weight v and function f,

$$\int (S(f))^2 \, v \, dx \le C \int |f(x)|^2 \, M_d(v) \, dx. \tag{3.3}$$

This will yield the result. How? Since the estimate we are seeking is purely local, we can, without loss of generality, assume that f's support is contained in I and that $\int_I f \, dx = 0$. In chapter 2 we proved inequality 2.15, which says that, if v is supported on a dyadic interval I and $\int_I v > 0$, then

$$c_1 \int_I M_d(v) \, dx \le \int_I v(x) \, \log(e + v(x)/v_I) \, dx \le c_2 \int_I M_d(v) \, dx,$$

where c_1 and c_2 are positive, absolute constants. Supposing we have 3.3, let I be a dyadic interval and set $E_\lambda = \{x \in I : \ S(f)(x) > \lambda\}$. Define $v = \chi_{E_\lambda}$. Then the left-hand side of 3.3 is bounded below by $\lambda^2|E_\lambda|$, while, because of our assumption about f's support, the right-hand side is bounded above by $C|E_\lambda| \log(e + |I|/|E_\lambda|)$. If $|E_\lambda| = 0$, we're done. Otherwise, after canceling, we get

$$\lambda^2 \le C \log(e + |I|/|E_\lambda|),$$

which, after some algebra, yields

$$|E_\lambda| \le C_1 \exp(-C_2\lambda^2)|I|.$$

So, how do we prove 3.3? For each integer k, let $\mathcal{F}^k$ denote the family of dyadic intervals I for which

$$2^k < \frac{1}{|I|} \int_I v \, dx \le 2^{k+1},$$

and define $D^k = \{x \in \mathbf{R} : M_d(v)(x) > 2^k\}$. We observe that $\cup\{I : I \in \mathcal{F}^k\} \subset D^k$, and that every dyadic I for which $\int_I v > 0$ lies in one and only one of the sets $\mathcal{F}^k$.

Now we write

$$\int (S(f))^2 \, v \, dx = \sum_I \frac{|\lambda_I|^2}{|I|} \int_I v \, dx$$

$$= \sum_k \sum_{I \in \mathcal{F}^k} |\lambda_I|^2 \frac{1}{|I|} \int_I v \, dx$$

$$\leq \sum_k 2^{k+1} \sum_{I \in \mathcal{F}^k} |\lambda_I|^2. \tag{3.4}$$

Now comes the trick. For any I, $\lambda_I = \langle f, h_{(I)} \rangle$. However, if $I \in \mathcal{F}^k$ then $I \subset D^k$, implying that $\lambda_I = \langle f\chi_{D^k}, h_{(I)} \rangle$. Therefore, the right-hand side of 3.4 is equal (respectively, less than or equal to):

$$\sum_k 2^{k+1} \sum_{I \in \mathcal{F}^k} |\langle f\chi_{D^k}, h_{(I)} \rangle|^2 \leq \sum_k 2^{k+1} \left(\int |f\chi_{D^k}|^2 \, dx \right)$$

$$= \int |f(x)|^2 \left(\sum_k 2^{k+1}\chi_{D^k} \right) dx$$

$$\leq 4 \int |f(x)|^2 \, M_d(v) \, dx,$$

because $\sum_k 2^{k+1}\chi_{D^k} \leq 4M_d(v)$. This finishes the proof of Theorem 3.1.

Proof of Theorem 3.2. The result will follow from a lemma.

Lemma 3.1. *Let g be supported in $[0,1)$ and satisfy $\int_0^1 g \, dx = 0$. Then:*

$$\int_0^1 \exp(g - \frac{1}{2}(S(g))^2) \, dx \leq 1. \tag{3.5}$$

To see how this applies to our problem, note that, by rescaling, it is enough to prove Theorem 3.2's exponential square estimate for $I = [0,1)$, under the hypotheses that f is supported in $[0,1)$ and satisfies $\int f \, dx = 0$. Under these assumptions, and with the additional one that $S(f) \leq 1$ almost everywhere, set $g = \lambda f$. Then 3.5 says

$$\int_0^1 \exp(\lambda f - \frac{1}{2}\lambda^2) \, dx \leq 1,$$

or

$$\int_0^1 \exp(\lambda f) \, dx \leq \exp(\frac{1}{2}\lambda^2),$$

from which Chebyshev's inequality yields:

$$|\{x \in [0,1): \ f(x) > \lambda\}| \leq \exp(-\frac{1}{2}\lambda^2).$$

Switching $-\lambda$ for λ gives the same inequality for $|\{x \in [0,1): \ f(x) < -\lambda\}|$. Combining the two, we get

$$|\{x \in [0,1): \ |f(x)| > \lambda\}| \leq 2\exp(-\frac{1}{2}\lambda^2),$$

which is what we want. So our problem reduces to showing Lemma 3.1.

The proof works by *induction*.

Define $g_0 = 0$ and, for $k > 0$,

$$g_k = g_{k-1} + \sum_{I: \ \ell(I)=2^{-k+1}} \lambda_I h_{(I)}.$$

Notice that g_k is simply what we get when we replace g by its averages over dyadic intervals of length 2^{-k}. We have implicitly assumed that $g \in L^1$. Therefore $g_k \to g$ and $S(g_k) \to S(g)$ almost everywhere as $k \to \infty$, and it is enough to prove 3.5 for g_k, since the general result will follow by Fatou's Lemma.

The case of $k = 0$ is trivial (both g_0 and $S(g_0)$ are 0). Now assume that

$$\int_0^1 \exp(g_k - \frac{1}{2}(S(g_k))^2) \, dx \leq 1.$$

Let $I \subset [0,1)$ be any dyadic interval of length 2^{-k}. It will be enough to show that

$$\int_I \exp(g_{k+1} - \frac{1}{2}(S(g_{k+1}))^2) \, dx \leq \int_I \exp(g_k - \frac{1}{2}(S(g_k))^2) \, dx.$$

But, since g_k and $S(g_k)$ are both constant across I, this will follow if we show that

$$\int_I \exp(g_{k+1} - g_k - \frac{1}{2}[(S(g_{k+1}))^2 - (S(g_k))^2]) \, dx \leq 1.$$

Let I_l and I_r denote I's left and right halves, respectively. The function $g_{k+1} - g_k$ has constant value—call it c—on I_l and value $-c$ on I_r. The function $(S(g_{k+1}))^2 - (S(g_k))^2$ has constant value c^2 on all of I. Thus, our problem reduces (last time!) to showing

$$\int_{I_l} \exp(c) \, dx + \int_{I_r} \exp(-c) \, dx \leq |I| \exp(c^2/2).$$

Transposing, this is the same as:

$$\frac{1}{|I|} \left[\int_{I_l} \exp(c) \, dx + \int_{I_r} \exp(-c) \, dx \right] \leq \exp(c^2/2). \tag{3.6}$$

However, an easy computation shows that the left-hand side of 3.6 is just $\cosh(c)$, which *is* $\leq \exp(c^2/2)$ (compare their power series). Theorem 3.2 is proved.

Remark. The reader might reasonably wonder why we didn't prove Theorem 3.2 by first showing that

$$\int |f(x)|^2 \, v \, dx \leq C \int (S(f))^2 \, M_d(v) \, dx \tag{3.7}$$

for all v and all appropriate f. The reason is simple: inequality 3.7 is false. The counterexample isn't hard to present. For $k = 0, 1, 2, \ldots$, let $I_k = [0, 2^{-k})$. Now let N be a large integer and define

$$v(x) = 2^N \chi_{I_N}(x)$$

$$f(x) = \sum_{k=0}^{N-1} \frac{1}{2^{k/2}(N-k)} h_{(I_k)}(x).$$

Then $|f(x)| = 1 + 1/2 + 1/3 + \cdots + 1/N \sim \log N$ on I_N, and so

$$\int |f(x)|^2 \, v \, dx \geq C(\log N)^2.$$

However, on $[2^{-k-1}, 2^{-k}]$ $(0 < k < N - 1)$, $M_d(v) \sim 2^k$ and

$$(S(f))^2 \leq \frac{1}{(N-k)^2} + \frac{1}{(N-k+1)^2} + \cdots + \frac{1}{N^2}$$
$$\leq C/(N-k),$$

implying $\int (S(f))^2 \, M_d(v) \, dx \leq C \log N$.

In the preceding chapter we devoted approximately equal space to inequalities of the form

$$\int (f^*(x))^p \, v \, dx \leq C \int (S(f))^p \, v \, dx$$

and those of the form

$$\int (S(f))^p \, v \, dx \leq C \int (f^*(x))^p \, v \, dx.$$

We are now forced to change that practice. We want to understand the two-weight problems

$$\int (f^*(x))^p \, v \, dx \leq C \int (S(f))^p \, w \, dx \tag{3.8}$$

and

$$\int (S(f))^p \, v \, dx \leq C \int (f^*(x))^p \, w \, dx. \tag{3.9}$$

At present, the theory of 3.8 is well-developed and (we think) well-understood; and the exponential-square result Theorem 3.2 plays a big role in this theory. But what I have just said is only partly true of 3.9. The basic L^2 inequality,

$$\int (S(f))^2\, v\, dx \le C \int |f(x)|^2\, M_d(v)\, dx,$$

can be generalized to L^p for $1 < p < 2$, yielding

$$\int (S(f))^p\, v\, dx \le C \int |f(x)|^p\, M_d(v)\, dx. \tag{3.10}$$

However, the proof makes no use of exponential-square estimates: it works by an interpolation argument like the one used to the prove the Hardy-Littlewood maximal theorem. If $p > 2$ then 3.10 fails; we show this at the end of the chapter. When p is large, substitutes for 3.10 are available, in which $M_d(v)$ is replaced by bigger maximal functions of v—such as iterations of $M_d(\cdot)$. However, the proofs of these results also make no use of exponential-square estimates. What they use is the theory of Orlicz spaces, which we will develop later in the book.

In order to maintain the flow of our narrative, we will *temporarily* shift our attention entirely to inequalities of the form 3.8. At the end of this chapter we will prove 3.10 for $1 < p < 2$, show that it fails for $p > 2$, and show an extension to $0 < p \le 1$. Unfortunately, the "large-p" analogues of 3.10—akin to what we will do with inequalities of the form 3.8—will have to wait until we have looked at Orlicz space theory.

So, given all that, our next order of business is to show how Theorem 3.2 lets us generalize the inequality

$$\int (f^*(x))^p\, v\, dx \le C \int (S(f))^p\, v\, dx$$

to two-weight settings

$$\int (f^*(x))^p\, v\, dx \le C \int (S(f))^p\, w\, dx \tag{3.11}$$

in which neither v nor w is assumed to belong to A_∞^d. We begin by introducing a useful functional.

Definition 3.1. *If v is a weight and I is a bounded interval, we define*

$$Y(I,v) = \begin{cases} \frac{1}{v(I)} \int_I v(x) \log(e + v(x)/v_I)\, dx & \text{if } v(I) > 0; \\ 1 & \text{if } v(I) = 0. \end{cases}$$

The Y-functional measures the extent to which v concentrates a lot of its mass on a small subset of I. For example, if we take $I = [0,1)$ and set, for $n \ge 0$,

$$v(x) = \begin{cases} 1 & \text{if } 0 \leq x \leq 2^{-n}; \\ 0 & \text{otherwise,} \end{cases}$$

then $Y(I, v) \sim n$.

The following theorem, which we mentioned in the last chapter, shows the intimate connection between the Y-functional and the A_∞^d property.

Theorem 3.3. *Let v be a weight. Then $v \in A_\infty^d$ if and only if there is a positive constant A such that, for all dyadic intervals I,*

$$\int_I M_d(v\chi_I) \, dx \leq Av(I). \tag{3.12}$$

Proof of Theorem 3.3. Suppose first that, for all dyadic intervals I,

$$\int_I M_d(v\chi_I) \, dx \leq Av(I).$$

By our inequality 2.15, this implies

$$\int_I v(x) \log(e + v(x)/v_I) \, dx \leq CAv(I) \tag{3.13}$$

for all $I \in \mathcal{D}$. Fix I, and, for $\lambda > 0$, define $E_\lambda = \{x \in I : v(x) > e^\lambda v_I\}$. By 3.13 and Chebyshev's inequality, we have

$$v(E_\lambda) \leq (A/\lambda)v(I). \tag{3.14}$$

Let $\epsilon > 0$ and take λ so large that $e^{-\lambda/2}$ and A/λ are both less than $\epsilon/4$. Now let $E \subset I$ and suppose $|E|/|I| < e^{-\lambda}$. Then:

$$v(E) \leq v(E \cap E_{\lambda/2}) + v(E \setminus E_{\lambda/2}).$$

By 3.14, the first term on the right is no bigger than $(2A/\lambda)v(I) \leq (\epsilon/2)v(I)$. The second term is less than or equal to

$$e^{\lambda/2}v_I|E| \leq e^{-\lambda/2}v(I) \leq (\epsilon/2)v(I).$$

Therefore, $v \in A_\infty^d$.

For the other direction, let $\delta > 0$ be so small that, for all dyadic I and measurable $E \subset I$, $|E|/|I| \leq \delta$ implies $v(E) \leq (1/2)v(I)$. Fix a dyadic interval I_0, which we can, without loss of generality, take to be $[0, 1)$, and suppose— also without loss of generality—that $\int_{I_0} v \, dx = 1$. Set $R = 2\delta^{-1}$, and, for $k = 1, 2, \ldots$, let $\{I_j^k\}_j$ be the maximal dyadic subintervals $I_j^k \subset I_0$ such that $v_{I_j^k} > R^k$. By maximality, each I_j^k satisfies

$$v_{I_j^k} \leq 2R^k$$

or

$$\int_{I_j^k} v\, dx \le 2R^k |I_j^k|.$$

Therefore, following the argument for the weak $(1,1)$ estimate for M_d,

$$\sum_{j': I_{j'}^{k+1} \subset I_j^k} |I_{j'}^{k+1}| \le \frac{2R^k}{R^{k+1}} |I_j^k|$$

$$= \delta |I_j^k|,$$

for all k and j. Because of our choice of R, this implies, for all k and j,

$$\sum_{j': I_{j'}^{k+1} \subset I_j^k} v(I_{j'}^{k+1}) \le (1/2) v(I_j^k), \tag{3.15}$$

and

$$\sum_{j': I_{j'}^{1} \subset I_0} v(I_{j'}^{1}) \le (1/2) v(I_0) = 1/2. \tag{3.16}$$

If we sum 3.15 over all j, for fixed $k \ge 1$, we get:,

$$\sum_{j'} v(I_{j'}^{k+1}) \le (1/2) \sum_{j} v(I_j^k),$$

because every $I_{j'}^{k+1}$ is contained in *some* I_j^k. We can continue this, going backward in k, to obtain, for each $k \ge 1$,

$$\sum_{j} v(I_j^k) \le (1/2) \sum_{j'} v(I_{j'}^{k-1})$$

$$\le (1/2)^2 \sum_{j'} v(I_{j'}^{k-2})$$

$$\le \cdots$$

$$\le 2^{-k} v(I_0) = 2^{-k}.$$

An argument we used in chapter 2 implies that $M_d(v\chi_{I_0}) \sim 1 + \sum_{k,j} R^k \chi_{I_j^k}$ on I_0, with approximate proportionality constants depending on R. Therefore:

$$\int_{I_0} M_d(v\chi_{I_0})\, dx \le C_R \left(1 + \int \left(\sum_{k,j} R^k \chi_{I_j^k}\right) dx\right)$$

$$\le C_R \left(1 + \sum_{k,j} R^k |I_j^k|\right)$$

$$\leq C_R \left(1 + \sum_{k,j} \int_{I_j^k} v(x)\, dx \right)$$

$$\leq C_R' \left(1 + \sum_{k} 2^{-k} \right)$$

$$\leq C_R' = C_R' v(I_0),$$

which was to be proved.

Theorem 3.3 implies the following *sufficient* condition for a weight v to belong to A_∞^d, which we mentioned in chapter 2.

Corollary 3.3. *If v is a weight and there is a positive constant A such that*

$$M_d(v) \leq Av \tag{3.17}$$

almost everywhere, then $v \in A_\infty^d$.

As we have seen, we can obtain weights satisfying 3.17 by applying the Rubio de Francia maximal function $\mathbf{M}_R$ to a weight w. Another way is by means of the following lemma, also mentioned in chapter 2.

Lemma 3.2. . *Let $0 < \beta < 1$. If v and $(M_d(v))^\beta$ are weights, then $(M_d(v))^\beta$ satisfies 3.17, with a constant A depending only on β.*

Proof of Lemma 3.2. Let $I = [0, 1)$ and suppose that $x_0 \in I$ is arbitrary. It will be enough to show that

$$\int_I (M_d(v)(x))^\beta\, dx \leq C_\beta (M_d(v)(x_0))^\beta,$$

because then, by rescaling and translating, we will have, for all dyadic intervals J and all points $y \in J$,

$$\frac{1}{|J|} \int_J (M_d(v)(x))^\beta\, dx \leq C_\beta (M_d(v)(y))^\beta,$$

which is 3.17.
Write

$$v = v_1 + v_2 \equiv v\chi_I + v\chi_{\mathbf{R}\setminus I}.$$

We will show

$$\int_I (M_d(v_1)(x))^\beta\, dx \leq C_\beta (M_d(v)(x_0))^\beta \tag{3.18}$$

and

$$\int_I (M_d(v_2)(x))^\beta\, dx \leq C_\beta (M_d(v)(x_0))^\beta. \tag{3.19}$$

For the first inequality 3.18, we use inequality 2.16. It directly gives us

$$\int_I (M_d(v_1)(x))^\beta \, dx \leq C_\beta \left(\int_I |v_1| \, dx \right)^\beta \leq C_\beta (M_d(v)(x_0))^\beta.$$

For the second, we note that $M_d(v_2)$ must be *constant* on I. This is because any dyadic interval that meets both I and v_2's support must contain I as a subset. This constant value—call it c—is less than or equal to the infimum of $M_d(v)$ on I, from which 3.19 now follows.

We have seen that Theorem 3.3 and Lemma 3.2 imply, for many weights v, the existence of weights w such that

$$\int (f^*(x))^p \, v \, dx \leq C \int (S(f))^p \, w \, dx \tag{3.20}$$

holds for all finite linear sums of Haar functions. Unfortunately, the w's so obtained tend to be rather singular (i.e., likely to blow up) maximal functions of v. We are about to use the Y-functional to find less singular w's, at least for some ranges of p. (Later we will able to do so for *all* p.)

A careful examination of the proof of Theorem 3.3 reveals the following very useful *quantitative* estimate: There is an absolute constant C such that, for all positive A and λ, if $Y(I, v) \leq A$ and $E \subset I$ satisfies $|E|/|I| < e^{-\lambda}$, then

$$v(E) \leq \frac{CA}{\lambda} v(I).$$

We shall use this fact in the following

Theorem 3.4. *Let $0 < p < \infty$. There is a positive constant C_p so that, for all weights v, if*

$$\sup_{I \in \mathcal{D}} Y(I, v) \leq A,$$

then, for all finite linear sums $f = \sum_I \lambda_I h_{(I)}$,

$$\int (f^*(x))^p \, v \, dx \leq C_p A^{p/2} \int (S(f))^p \, v \, dx.$$

Proof of Theorem 3.4. Write $\{x \in \mathbf{R} : f^*(x) > \lambda\} = \cup_i I_i^\lambda$, where the I_i^λ are the usual maximal dyadic intervals. For every $\gamma > 0$ and I_i^λ, we have

$$|\{x \in I_i^\lambda : f^*(x) > 2\lambda, \, S(f)(x) \leq \gamma\lambda\}| \leq C_1 \exp(-C_2 \gamma^{-2})|I_i^\lambda|.$$

Because of our hypothesis on v, this implies

$$v(\{x \in I_i^\lambda : f^*(x) > 2\lambda, \, S(f)(x) \leq \gamma\lambda\}) \leq (CA\gamma^2)v(I_i^\lambda).$$

If we choose γ such that $CA\gamma^2 = 2^{-p-1}$, we will get

$$\int (f^*(x))^p \, v \, dx \leq 2^{p+1} \gamma^{-p} \int (S(f))^p \, v \, dx.$$

(The reader might want to go through the original good-λ inequality proof again and check this.) But

$$\gamma = \frac{1}{2^{(p+1)/2} \sqrt{CA}},$$

and that implies our estimate.

In order to squeeze the most out of the Y-functional, we will need a more severely "localized" form of Theorem 3.4.

Theorem 3.5. *Let v be a weight and let $\mathcal{F}$ be a finite family of dyadic intervals such that, for all $I \in \mathcal{F}$, $Y(I, v) \leq A$. If $f = \sum_{I \in \mathcal{F}} \lambda_I h_{(I)}$ then, for all $0 < p < \infty$,*

$$\int (f^*(x))^p \, v \, dx \leq C_p A^{p/2} \int (S(f))^p \, v \, dx,$$

where the constant C_p only depends on p.

Remark. The proof of this theorem might at first appear trivial, since the only intervals that "count" are those in $\mathcal{F}$, which satisfy the good estimate $Y(I, v) \leq A$. However, things are not quite so simple, because we need to prove our good-λ estimate on the stopping intervals I_i^λ, *each of which is a right or left half of some $I \in \mathcal{F}$, and we don't know anything about those.*

Proof of Theorem 3.5. Let I_i^λ be a maximal dyadic interval such that $|f_{I_i^\lambda}| > \lambda$. As always, we can assume that $|f_{I_i^\lambda}| \leq (1.1)\lambda$; and we do so. Let $\{K_j\}$ be the family of *maximal* $I \in \mathcal{F}$ which are also subsets of I_i^λ. We "observe" that if $J \subset I_i^\lambda$ is a dyadic interval that is not properly contained in any K_j, then $f_J = f_{I_i^\lambda}$. *This observation is the key to the proof.* Why is it true? Note that, for any dyadic interval L,

$$f_L = \left(\sum_{\substack{I \in \mathcal{F} \\ I \not\subset L}} \lambda_I h_{(I)} \right)_L,$$

because the terms $\lambda_I h_{(I)}$ with $I \subset L$ integrate to 0. Therefore,

$$f_J - f_{I_i^\lambda} = \left(\sum_{\substack{I \in \mathcal{F}: I \subset I_i^\lambda \\ I \not\subset J}} \lambda_I h_{(I)} \right)_J. \tag{3.21}$$

However, because of our assumption on J, the sum in 3.21 is empty. (Remember: we are using '$\subset$' to mean "subset," not "proper subset.") So the observation is true.

Because of the observation, we can now state that

$$v\left(\{x \in I_i^\lambda : \ f^*(x) > 2\lambda, \ S(f)(x) \le \gamma\lambda\}\right)$$

$$\le v\left(\{x \in I_i^\lambda : \ (f - f_{I_i^\lambda})^*(x) > .9\lambda, \ S(f)(x) \le \gamma\lambda\}\right)$$

$$= \sum_j v\left(\{x \in K_j : \ (f - f_{K_j})^*(x) > .9\lambda, \ S(f)(x) \le \gamma\lambda\}\right).$$

But each K_j satisfies $Y(K_j, v) \le A$. So, by choosing, as before, $\gamma = (2^{(p+1)/2}\sqrt{CA})^{-1}$, we get

$$\sum_j v\left(\{x \in K_j : \ (f - f_{K_j})^*(x) > .9\lambda, \ S(f)(x) \le \gamma\lambda\}\right)$$

$$\le 2^{-p-1} \sum_j v(K_j)$$

$$\le 2^{-p-1} v(I_i^\lambda),$$

and that does it!

Remark. Let us observe (what we've already seen) that

$$\int (S(f))^2 \, v \, dx = \sum_I \frac{|\lambda_I|^2}{|I|} \int_I v \, dx.$$

This will be useful shortly.

Before considering applications of Theorem 3.5 to the two-weight problem,

$$\int (f^*(x))^p \, v \, dx \le C \int (S(f))^p \, w \, dx$$

we note in passing an easy generalization. For $\eta > 0$, define

$$Y_\eta(I, v) = \begin{cases} \frac{1}{v(I)} \int_I v(x) \left(\log(e + v(x)/v_I)\right)^\eta dx & \text{if } v(I) > 0; \\ 1 & \text{if } v(I) = 0. \end{cases} \tag{3.22}$$

The proof of the next theorem closely follows that of Theorem 3.5, and is left to the reader.

Theorem 3.6. *Let v be a weight and let $\mathcal{F}$ be a finite family of dyadic intervals such that $Y_\eta(I, v) \le A$ for all $I \in \mathcal{F}$. If $0 < p < \infty$ and $f = \sum_{I \in \mathcal{F}} \lambda_I h_{(I)}$ then*

$$\int (f^*(x))^p \, v \, dx \le C A^{p/(2\eta)} \int (S(f))^p \, v \, dx,$$

where the constant C only depends on p and η.

The proof of the next theorem will be a model for the L^p arguments which follow it.

Theorem 3.7. *Let v be a weight and let $\tau > 1$. There is a $C = C(\tau)$ such that, if $f = \sum_I \lambda_I h_{(I)}$ is any finite linear sum, then*

$$\int (f^*(x))^2 \, v \, dx \le C \sum_I \frac{|\lambda_I|^2}{|I|} \int_I v(x) \, (\log(e + v(x)/v_I))^\tau \, dx.$$

Proof of Theorem 3.7. By an easy limiting argument, we can assume that $Y(I, v)$ is finite for all I in our sum. For $k = 0, 1, 2, \ldots$, define

$$\mathcal{F}_k = \{I : 2^k \le Y(I, v) < 2^{k+1}\}$$

and

$$f_{(k)} \equiv \sum_{I : I \in \mathcal{F}_k} \lambda_I h_{(I)}.$$

Now set $\delta = \tau - 1 > 0$ and write:

$$\int (f^*(x))^2 \, v \, dx \le \int \left(\sum_k (f_{(k)})^*(x) \right)^2 v \, dx$$

$$\le C_\tau \sum_k 2^{k\delta} \int (f_{(k)}(x))^2 \, v \, dx$$

$$\le C_\tau \sum_k 2^{k\delta} 2^k \int (S(f_{(k)}))^2 \, v \, dx$$

$$= C_\tau \sum_k 2^{k\tau} \sum_{I \in \mathcal{F}_k} \frac{|\lambda_I|^2}{|I|} \int_I v \, dx, \qquad (3.23)$$

where the next-to-last line follows from Theorem 3.5 (applied to each $f_{(k)}$) and the last line is a consequence of our remark above.

For every $I \in \mathcal{F}_k$,

$$2^{k\tau} v(I) \, dx \le v(I) \left(\frac{1}{v(I)} \int_I v(x) \, \log(e + v(x)/v_I) \, dx \right)^\tau$$

$$\le v(I) \frac{1}{v(I)} \int_I v(x) \, (\log(e + v(x)/v_I))^\tau \, dx$$

$$\le \int_I v(x) \, (\log(e + v(x)/v_I))^\tau \, dx.$$

Therefore, the right-hand side of 3.23 is less than or equal to

$$C_\tau \sum_k \sum_{I \in \mathcal{F}_k} \frac{|\lambda_I|^2}{|I|} \int_I v(x) \, (\log(e + v(x)/v_I))^\tau \, dx$$

$$\le C_\tau \sum_I \frac{|\lambda_I|^2}{|I|} \int_I v(x) \, (\log(e + v(x)/v_I))^\tau \, dx$$

which is the result we sought.

We immediately get this sufficient condition for the $L^2 - L^2$ two-weight inequality.

Corollary 3.4. *Let $\tau > 1$, and let v and w be weights such that*

$$\int_I v(x) \left(\log(e + v(x)/v_I)\right)^\tau dx \leq \int_I w(x)\, dx$$

for all $I \in \mathcal{D}$. Then

$$\int (f^*(x))^2\, v\, dx \leq C_\tau \int (S(f))^2\, w\, dx$$

for all finite sums $f = \sum_I \lambda_I h_{(I)}$.

Things are relatively simple in the L^2 case, because (as we've already noted),

$$\int (S(f))^2\, v\, dx = \sum_I \frac{|\lambda_I|^2}{|I|} v(I).$$

In dealing with L^p, $p \neq 2$, we shall have to be trickier. Temporarily fix $0 < p < \infty$ and let $f = \sum_I \lambda_I h_{(I)}$ be a finite linear sum. For every $I \in \mathcal{D}$, define

$$c(p, I) \equiv \left(\sum_{J:\, I \subset J} \frac{|\lambda_J|^2}{|J|} \right)^{p/2} - \left(\sum_{\substack{J:\, I \subset J \\ J \neq I}} \frac{|\lambda_J|^2}{|J|} \right)^{p/2}.$$

The non-negative numbers $\{c(p, I)\}$ have two valuable properties.
First, $c(p, I) \neq 0$ if and only if $\lambda_I \neq 0$.
Second, $(S(f)(x))^p = \sum_{I:x \in I} c(p, I)$. (It's a telescoping sum.)
We now have the tools to prove a pretty good sufficient condition for the two-weight inequality 3.8.

Theorem 3.8. *Let $0 < p < \infty$ and let $\tau > p/2$. Let v and w be weights such that*

$$\int_I v(x) \left(\log(e + v(x)/v_I)\right)^\tau dx \leq \int_I w(x)\, dx$$

for all $I \in \mathcal{D}$. Then, for all finite sums $f = \sum_I \lambda_I h_{(I)}$,

$$\int (f^*(x))^p\, v\, dx \leq C \int (S(f))^p\, w\, dx,$$

where the constant C only depends on p and τ.

Proof of Theorem 3.8. The proofs for $p > 2$ and $p < 2$ are slightly different, but since the one for $p > 2$ more closely resembles those of Theorem 3.7 and its corollary, we'll do it first.

As before, we may assume that $Y(I, v) < \infty$ for all I. Define $\mathcal{F}_k$ and $f_{(k)}$ as we did earlier, and set $\epsilon = \tau - p/2 > 0$. Write:

$$\int \left(\sum_k (f^*_{(k)}(x)) \right)^p v \, dx \leq C \sum_k 2^{k\epsilon} \int (f^*_{(k)}(x))^p \, v \, dx$$

$$\leq C \sum_k 2^{k\epsilon} 2^{kp/2} \int (S(f_{(k)}))^p \, v \, dx$$

$$= C \sum_k 2^{k(\epsilon+p/2)} \int \left(\sum_{I : x \in I \in \mathcal{F}_k} c(p, I) \right) v \, dx$$

$$= C \sum_k \sum_{I : I \in \mathcal{F}_k} c(p, I) 2^{k(\epsilon+p/2)} v(I).$$

However, arguing exactly as in the proof of Theorem 3.7 (recall that $\tau = \epsilon + p/2$),

$$2^{k(\epsilon+p/2)} v(I) \leq \int_I v(x) \left(\log(e + v(x)/v_I) \right)^\tau dx$$

$$\leq w(I).$$

Thus:

$$\sum_k \sum_{I : I \in \mathcal{F}_k} c(p, I) 2^{k(\epsilon+p/2)} v(I) \leq C \sum_k \int \left(\sum_{I : x \in I \in \mathcal{F}_k} c(p, I) \right) w \, dx$$

$$= C \sum_k \int (S(f_{(k)}))^p \, w \, dx$$

$$= C \int \left(\sum_k (S(f_{(k)}))^p \right) w \, dx$$

$$\leq C \int (S(f))^p \, w \, dx,$$

where the last line follows because $(S(\sum_k f_{(k)}))^p \geq \sum_k (S(f_{(k)}))^p$ when $p/2 \geq 1$. This finishes the "$p > 2$" half of the proof.

Now take $p < 2$ and let $p/2 < \eta < \tau$. The reader should recall the definition of $Y_\eta(I, v)$ from 3.22. We assume that $Y_\eta(I, v)$ is finite for all I. Now define, for $k = 0, 1, 2, \ldots$,

$$\mathcal{F}_{k,\eta} = \{ I : \ 2^k \leq Y_\eta(I, v) < 2^{k+1} \}$$

$$f_{(k,\eta)} = \sum_{I \in \mathcal{F}_{k,\eta}} \lambda_I h_{(I)}.$$

Let $\epsilon = (1/2)(\tau - p/2)/\eta > 0$ and (much as before) write:

$$\int (f^*(x))^p \, v \, dx \leq C \sum_k 2^{k\epsilon} \int (f^*_{(k,\eta)}(x))^p \, v \, dx$$

$$\leq C \sum_k 2^{k\epsilon} 2^{kp/(2\eta)} \int (S(f_{(k)}))^p \, v \, dx$$

$$= C \sum_k 2^{k(\epsilon + p/(2\eta))} \int \left(\sum_{I : x \in I \in \mathcal{F}_{k,\eta}} c(p, I) \right) v \, dx$$

$$= C \sum_k \sum_{I : I \in \mathcal{F}_{k,\eta}} c(p, I) 2^{k(\epsilon + p/(2\eta))} v(I).$$

But, for $I \in \mathcal{F}_{k,\eta}$,

$$2^k \leq \frac{1}{v(I)} \int_I v(x) \left(\log(e + v(x)/v_I) \right)^\eta dx$$

$$\leq \left(\frac{1}{v(I)} \int_I v(x) \left(\log(e + v(x)/v_I) \right)^\tau dx \right)^{\eta/\tau}$$

$$\leq \left(\frac{w(I)}{v(I)} \right)^{\eta/\tau},$$

which implies

$$v(I) \leq 2^{-k\tau/\eta} w(I).$$

When we substitute this estimate in, we get:

$$\sum_k \sum_{I : I \in \mathcal{F}_{k,\eta}} c(p, I) 2^{k(\epsilon + p/(2\eta))} v(I) \leq \sum_k \sum_{I : I \in \mathcal{F}_{k,\eta}} c(p, I) 2^{k(\epsilon + p/(2\eta))} 2^{-k\tau/\eta} w(I)$$

$$= C \sum_k 2^{-k\epsilon} \sum_{I : I \in \mathcal{F}_{k,\eta}} c(p, I) w(I)$$

$$= C \sum_k 2^{-k\epsilon} \int \left(\sum_{I : x \in I \in \mathcal{F}_{k,\eta}} c(p, I) \right) w \, dx$$

$$= C \sum_k 2^{-k\epsilon} \int (S(f_{(k)}))^p \, w \, dx$$

$$\leq C \int (S(f))^p \, w \, dx,$$

where this time we *need* that extra ϵ at the end. Theorem 3.8 is proved.

Theorem 3.8 immediately suggests a question: Given a positive number η and a weight v, what weights w satisfy

$$\int_I v(x) \left(\log(e + v(x)/v_I) \right)^\eta dx \leq \int_I w(x) \, dx \tag{3.24}$$

for all $I \in \mathcal{D}$? We have seen that

$$\int_I v(x) \log(e + v(x)/v_I) \, dx \sim \int_I M_d(\chi_I v) \, dx;$$

so, if $\eta \leq 1$, we may take $w = c M_d(v)$, with c some positive constant. It turns out that, for any positive integer k,

$$\int_I v(x)\,(\log(e + v(x)/v_I))^k\,dx \sim \int_I M_d^k(\chi_I v)\,dx, \tag{3.25}$$

where M_d^k denotes the k-fold iteration of the operator M_d. This approximate equivalence is well-known; we will give a proof of it in our chapter on Orlicz spaces. Granting the result, we can see that one candidate w for 3.24 is $c_\eta M_d^{[\eta]+1} v$, where $[\eta]$ is the greatest integer in η. We get:

Corollary 3.5. *Let* $0 < p < \infty$. *There is a positive constant* $C = C(p)$ *such that, for all weights* v *and all finite sums* $f = \sum_I \lambda_I h_{(I)}$,

$$\int (f^*(x))^p\,v\,dx \le C \int (S(f))^p\,M_d^{[p/2]+1}(v)\,dx.$$

Note that, if $p < 2$, Corollary 3.5 yields:

$$\int (f^*(x))^p\,v\,dx \le C \int (S(f))^p\,M_d(v)\,dx. \tag{3.26}$$

The reader will recall that 3.26 is *false* for $p = 2$. This fact—in light of Theorems 3.1 and 3.2 and inequality 3.3—seems to say that, while the control that $S(f)$ exercises over f and the control that f exercises over $S(f)$ are essentially equivalent, f's control over $S(f)$ is just a hair's breadth tighter.

For large k, the function $M_d^k(v)$ is an unsatisfactory majorant of v, for two reasons: 1) it's hard to estimate (it requires us to apply the M_d operator repeatedly); 2) it's sloppy—i.e., much too big—when η is not an integer. When we study Orlicz spaces, and just about the same time we see a proof of 3.25, we will learn about a family of majorants which avoid both of these problems.

Mini-Appendix I: The Necessity of the A_∞^d Condition

We will show that the inequality

$$\int_{\mathbf{R}} |f(x)|^2\,v\,dx \le C \int_{\mathbf{R}} (S(f))^2\,v\,dx$$

cannot hold if $v \notin A_\infty^d$. Recall that if $I \in \mathcal{D}$ and v is a weight,

$$Y(I, v) \sim \begin{cases} \int_I M_d(\chi_I v)\,dx / \int_I v\,dx & \text{if } v(I) > 0; \\ 1 & \text{otherwise.} \end{cases}$$

We have seen that if

$$\sup_I Y(I, v) = A < \infty$$

then

$$\int |f(x)|^2\,v\,dx \le CA \int (S(f))^2\,v\,dx, \tag{3.27}$$

valid for all finite linear sums of Haar functions, for some constant C independent of f and v. Now we will show that the bound CA in 3.27 is essentially sharp. It is sharp in a strong sense. It is not merely the case that, for every $A \geq 1$, we can find a weight v and a finite linear sum of Haar functions, f, such that

$$\sup_I Y(I, v) = A$$

and

$$\int |f(x)|^2 \, v \, dx \geq cA \int (S(f))^2 \, v \, dx,$$

where $c > 0$ does not depend on f and v. Instead, we have

Theorem 3.9. *There is a constant $c > 0$ such that, for all dyadic intervals J and all weights v, if $Y(J, v) > A > 1$, then we can find a function f, a finite linear sum of Haar functions, supported in J, satisfying $\int_J f \, dx = 0$, and such that*

$$\int_J |f(x)|^2 \, v \, dx \geq cA \int (S(f))^2 \, v \, dx.$$

Proof of Theorem 3.9. The proof depends on a construction and a few observations. Without loss of generality, we assume that $J = [0, 1)$, $v(J) = 1$, and A is very large (but finite). For $R = 10$ and $k \geq 1$, we let $\{I_j^k\}_j$ be the maximal dyadic subintervals of J such that $v_{I_j^k} > R^k$. Let us set $E^k \equiv \cup_j I_j^k$ for $k \geq 1$. Our value of R ensures that $|E^1| \leq 1/3$ and, for all I_j^k,

$$|E^{k+1} \cap I_j^k| \leq (1/3)|I_j^k|.$$

By our previous work, we know that

$$\int_J v(x) \left(\sum_{k,j} \chi_{I_j^k}(x) \right) \, dx \geq cA,$$

where c is an absolute constant[1]. We will build a (possibly unbounded) f, supported in J, with integral 0, and satisfying

$$|f(x)| \geq (1/2) \sum_{k,j} \chi_{I_j^k}(x).$$

But its square function $S(f)$ will satisfy

$$\int (S(f))^2 \, v \, dx \leq C \int_J v(x) \left(\sum_{k,j} \chi_{I_j^k}(x) \right) \, dx.$$

This is nearly what we want. Set $\psi(x) = \sum_{k,j} \chi_{I_j^k}(x)$. Then

---

[1] It depends on R, but we've fixed that.

$$\int_J |f(x)|^2 \, v \, dx \geq (1/4) \int_J (\psi(x))^2 \, v \, dx$$

while

$$\int (S(f))^2 \, v \, dx \leq C \int_J \psi(x) \, v(x) \, dx;$$

implying

$$\frac{\int_J |f(x)|^2 \, v \, dx}{\int (S(f))^2 \, v \, dx} \geq c \frac{\int_J (\psi(x))^2 \, v \, dx}{\int_J \psi(x) \, v(x) \, dx}$$

$$\geq c \int_J \psi(x) \, v(x) \, dx \geq cA,$$

where the next-to-last inequality follows by the Cauchy-Schwarz inequality and the normalization of v. A trivial approximation argument will then yield the f we seek.

Define, for $x \in J$,

$$\phi_0(x) = \begin{cases} 1 & \text{if } x \in E^1; \\ -|E^1|/|J \setminus E^1| & \text{if } x \notin E^1. \end{cases}$$

If $k \geq 1$, we "relativize" the definition of ϕ_0 to I_j^k by setting

$$\phi_{k,j}(x) = \begin{cases} 1 & \text{if } x \in E^{k+1} \cap I_j^k; \\ -|E^{k+1} \cap I_j^k|/|I_j^k \setminus E^{k+1}| & \text{if } x \in I_j^k \setminus E^{k+1}; \\ 0 & \text{otherwise.} \end{cases}$$

The functions $\phi_{k,j}$ all have supports contained in I_j^k, have values lying between 1 and $-1/2$ (because of our choice of R), and satisfy $\int_{I_j^k} \phi_{k,j} \, dx = 0$. Similarly, ϕ_0 is supported in J, has absolute value ≤ 1, and satisfies $\int_J \phi_0 \, dx = 0$. We set

$$f(x) \equiv \phi_0(x) + \sum_{k,j} \phi_{k,j}(x).$$

To see that $|f(x)| \geq (1/2) \sum_{k,j} \chi_{I_j^k}(x)$, suppose that $x \in I_j^l \setminus E^{l+1}$. Then, for some indices $j_1, j_2, \ldots,$

$$f(x) = \phi_0(x) + \phi_{1,j_1}(x) + \phi_{2,j_2}(x) + \cdots + \phi_{l-1,j_{l-1}}(x) + \phi_{l,j_l}(x)$$
$$\geq 1 + 1 + 1 + \cdots + 1 - (1/2)$$
$$\geq (1/2) \left(\chi_{1,j_1}(x) + \chi_{2,j_2}(x) + \cdots + \chi_{l-1,j_{l-1}}(x) + \chi_{l,j_l}(x) \right),$$

because all the summands defining $f(x)$ equal 1, except possibly the very last, and that one is no smaller than $-1/2$.

To establish our bound for $S(f)$, we look at $\langle f, h_{(I)} \rangle$, where $I \subset J$. If I is not contained in any I_j^k then $h_{(I)}$ is orthogonal to all of the $\phi_{k,j}$'s, and $\langle f, h_{(I)} \rangle = \langle \phi_0, h_{(I)} \rangle$. Otherwise, there is a unique I_j^k such that $I \subset I_j^k$ but

$I \not\subset \cup_{j'} I_{j'}^{k+1}$. For this I we will have $\langle f, h_{(I)} \rangle = \langle \phi_{k,j}, h_{(I)} \rangle$: the inner product $\langle \phi_{k',j'}, h_{(I)} \rangle$ will be zero for all other $\phi_{k',j'}$, because either I and $I_{j'}^{k'}$ are disjoint, or $h_{(I)}$ integrates to 0 across an interval on which $\phi_{k',j'}$ is constant, or the reverse happens. Therefore

$$(S(f))^2 = (S(\phi_0))^2 + \sum_{k,j} (S(\phi_{k,j}))^2,$$

and

$$\int (S(f))^2 \, v \, dx = \int_J (S(\phi_0))^2 \, v \, dx + \sum_{k,j} \int_{I_j^k} (S(\phi_{k,j}))^2 \, v \, dx.$$

We finish the construction by showing

$$\int_J (S(\phi_0))^2 \, v \, dx \le C v(J)$$

and, for each k and j,

$$\int_{I_j^k} (S(\phi_{k,j}))^2 \, v \, dx \le C v(I_j^k).$$

Since the arguments are very similar, we will only show the first of these inequalities.

Let $\mathcal{F} = \{I \in \mathcal{D} : I \subset J, \ I \not\subset \cup_j I_j^1\}$. The two important things to notice are that each of these intervals satisfies $v_I \le R$ and

$$(S(\phi_0))^2 = \sum_{I \in \mathcal{F}} \frac{|\langle \phi_0, h_{(I)} \rangle|^2}{|I|} \chi_I,$$

because only the I's in $\mathcal{F}$ yield non-zero inner products. Therefore

$$\int_J (S(\phi_0))^2 \, v \, dx = \sum_{I \in \mathcal{F}} \frac{|\langle \phi_0, h_{(I)} \rangle|^2}{|I|} v(I)$$
$$\le R \sum_{I \in \mathcal{F}} |\langle \phi_0, h_{(I)} \rangle|^2$$
$$= R \int_J |\phi_0(x)|^2 \, dx$$
$$\le R = 10 v(J),$$

which is what we wanted.

Exercise 14.1 outlines a proof that the dyadic doubling property (Definition 2.5) does not imply the dyadic A_∞ property. Therefore, dyadic doubling in a weight v is not sufficient for the inequality

$$\int_{\mathbf{R}} |f(x)|^2 \, v \, dx \le C \int_{\mathbf{R}} (S(f))^2 \, v \, dx$$

to hold.

Mini-Appendix II: Going the Other Way

We will show to what extent inequality 3.3 can be extended to p's below 2—leaving aside, for the time being, the problem of extending it to larger p's. We will need the following lemma:

Lemma 3.3. *If v is a weight and f is locally integrable then, for all $\lambda > 0$,*

$$v\left(\{x : \ M_d(f)(x) > \lambda\}\right) \leq \frac{C}{\lambda} \int |f| \, M_d(v) \, dx.$$

Proof of Lemma 3.3. We first assume that $f \in L^1$. Let $\{I_j\}$ be the maximal dyadic intervals such that

$$\frac{1}{|I_j|} \int_{I_j} |f| \, dt > \lambda.$$

By maximality,

$$\frac{1}{|I_j|} \int_{I_j} |f| \, dt \leq 2\lambda.$$

Therefore,

$$|I_j| \sim \frac{1}{\lambda} \int_{I_j} |f| \, dt$$

for each j. We need to estimate $\sum_j v(I_j)$. This is the same as

$$\sum_j \frac{v(I_j)}{|I_j|} |I_j|,$$

which is bounded by a constant times

$$\sum_j \frac{v(I_j)}{|I_j|} \frac{1}{\lambda} \int_{I_j} |f| \, dt.$$

However, $M_d(v) \geq \frac{v(I_j)}{|I_j|}$ on all of I_j. Therefore,

$$\sum_j \frac{v(I_j)}{|I_j|} \frac{1}{\lambda} \int_{I_j} |f| \, dt \leq \sum_j \frac{1}{\lambda} \int_{I_j} |f| \, M_d(v) \, dt$$

$$\leq \frac{1}{\lambda} \int |f| \, M_d(v) \, dt,$$

proving the lemma for integrable f. For the general case, replace f by $f_N \equiv f\chi_{[-N,N]}$ and let $N \to \infty$. (We will leave the details of this to the reader, but give him the crucial hint that $M_d(f_N) \nearrow M_d(f)$ by Monotone Convergence.)

Lemma 3.3 says that M_d is weak type $(1,1)$ when mapping from $L^1(M_d(v))$ to $L^1(v)$, with a constant independent of v. Since M_d is trivially weak type (∞, ∞), Theorem 2.8 implies

Corollary 3.6. *If $1 < p < \infty$, there is a C_p such that*

$$\int (M_d(f))^p \, v \, dx \leq C_p \int |f(x)|^p \, M_d(v) \, dx$$

for all functions f and weights v.

We now recall our condition 1.4 from chapter 1: for every $\epsilon > 0$, there is an R such that, if I is any interval with $\ell(I) > R$, then

$$\frac{1}{|I|} \int_I |f| \, dt < \epsilon.$$

Theorem 3.10. *If $1 < p \leq 2$ then*

$$\int (S(f))^p \, v \, dx \leq C_p \int |f(x)|^p \, M_d(v) \, dx,$$

for all weights v, and all functions f satisfying 1.4, where the constant C_p only depends on p.

Proof of Theorem 3.10. We know that

$$\int (S(f))^2 \, v \, dx \leq C \int |f(x)|^2 \, M_d(v) \, dx,$$

which amounts to saying that $S(\cdot)$ maps boundedly from $L^2(M_d(v))$ to $L^2(v)$. By Theorem 2.8, all we need to do is to show that it is also weak type $(1,1)$.

Let $\lambda > 0$, and let I_j^λ be the maximal dyadic intervals such that

$$\frac{1}{|I_j^\lambda|} \int_{I_j^\lambda} |f| \, dt > \lambda;$$

such intervals exist by 1.4. We write $f = g + b$, where g and b are the "good" and "bad" functions corresponding to the intervals $\{I_j^\lambda\}$. In other words,

$$g(x) = \begin{cases} f(x) & \text{if } x \notin \cup_j I_j^\lambda; \\ f_{I_j^\lambda} & \text{if } x \in I_j^\lambda; \end{cases}$$

and $b = \sum b_j$, where

$$b_j = (f - f_{I_j^\lambda}) \chi_{I_j^\lambda}.$$

We observe that $|g| \leq 2\lambda$ everywhere, and that

$$|g| \leq \frac{1}{|I_j^\lambda|} \int_{I_j^\lambda} |f| \, dt$$

on each I_j^λ. We also observe that the support of $S(b)$ is contained entirely inside $\cup_j I_j^\lambda$ (this follows from equations 2.4 and 2.5).

We recall that $\Omega_\lambda \equiv \cup_j I_j^\lambda$ is the set where $M_d(f) > \lambda$. Lemma 3.3 says that this set has v-measure no larger than

$$\frac{C}{\lambda} \int |f| \, M_d(v) \, dx.$$

Therefore it is enough to show that

$$v\left(\{x \notin \Omega_\lambda : \; S(f)(x) > \lambda\}\right) \le \frac{C}{\lambda} \int |f| \, M_d(v) \, dx.$$

But $S(f) \le S(g) + S(b)$, and $S(b) \equiv 0$ on the complement of Ω_λ. Therefore it suffices to show

$$v\left(\{x \notin \Omega_\lambda : \; S(g)(x) > \lambda\}\right) \le \frac{C}{\lambda} \int |f| \, M_d(v) \, dx, \qquad (3.28)$$

and that is what we will do.

The proof of 3.28 requires a simple lemma, whose proof we leave to the reader.

Lemma 3.4. *If v is any weight and J is any dyadic interval,*

$$\sup_{x \in J} M_d(v\chi_{\mathbf{R}\setminus J})(x) \le C \inf_{x \in J} M_d(v)(x).$$

Given this, we use our L^2 inequality:

$$\int_{\mathbf{R}\setminus\Omega_\lambda} (S(g))^2 \, v \, dx = \int (S(g))^2 \, (v\chi_{\mathbf{R}\setminus\Omega_\lambda}) \, dx$$

$$\le C \int |g|^2 \, M_d((v\chi_{\mathbf{R}\setminus\Omega_\lambda})) \, dx$$

$$= C \int_{\mathbf{R}\setminus\Omega_\lambda} |f|^2 \, M_d(v) \, dx + \sum_j \int_{I_j^\lambda} |g|^2 \, M_d((v\chi_{\mathbf{R}\setminus\Omega_\lambda})) \, dx$$

$$\equiv (I) + (II).$$

Since $|f| \le \lambda$ almost-everywhere off Ω_λ, quantity (I) is no larger than

$$C\lambda \int |f| \, M_d(v) \, dx.$$

Each of the summands in (II) is no bigger than

$$C\lambda \left(\frac{1}{|I_j^\lambda|} \int_{I_j^\lambda} |f| \, dt \right) \int_{I_j^\lambda} M_d((v\chi_{\mathbf{R}\setminus\Omega_\lambda})) \, dt,$$

which, by Lemma 3.4, is no larger than

$$C\lambda \int_{I_j^\lambda} |f| \, M_d(v) \, dt.$$

These facts imply that $(I) + (II)$ is no bigger than a constant times

$$C\lambda \int |f| \, M_d(v) \, dx.$$

Inequality 3.28 now follows from Chebyshev's inequality (dividing both sides by λ^2).

Theorem 3.10 fails for $p \leq 1$. If $f = \chi_{[0,1)}$ and $v \equiv 1$, then

$$\int |f(x)|^p \, M_d(v) \, dx = 1.$$

When $2^k \leq x < 2^{k+1}$ ($k \geq 0$), we have $S(f) = 2^{-k}$, implying that

$$\int (S(f))^p \, M_d(v) \, dx = \infty$$

if $0 < p \leq 1$.

However, for small p, a substitute inequality is true.

Theorem 3.11. *If $0 < p \leq 1$ then*

$$\int (S(f))^p \, v \, dx \leq C \int (f^*)^p \, M_d(v) \, dx,$$

where the constant C only depends on p.

Proof of Theorem 3.11. We can assume that v has bounded support; the general result will then follow by Monotone Convergence. For $k = 0, \pm 1, \pm 2, \ldots$, let $\{I_j^k\}$ be the maximal dyadic intervals such that $v_{I_j^k} > 2^k$. Define E_j^k to be the collection of $I \in \mathcal{D}$ which are subsets of I_j^k but are not contained in any $I_{j'}^{k+1}$. Write:

$$f_{(k,j)} = \sum_{I:I \in E_j^k} \lambda_I(f) h_{(I)}.$$

Then $(S(f))^p \leq \sum_{k,j} (S(f_{(k,j)}))^p$, because $p \leq 2$. Also, each $S(f_{(k,j)})$ is supported in I_j^k. Therefore

$$\int (S(f))^p \, v \, dx \leq \sum_{k,j} \int_{I_j^k} (S(f_{(k,j)}))^p \, v \, dx.$$

We will be done if we can show that

$$\int_{I_j^k} (S(f_{(k,j)}))^p \, v \, dx \leq C_p 2^k \int_{I_j^k} (f_{(k,j)}^*)^p \, dx, \tag{3.29}$$

because, summing on k and j will then yield:

$$\int (S(f))^p \, v \, dx \leq C_p \sum_{k,j} \int (f_{(k,j)}^*)^p \left(2^k \chi_{I_j^k} \right) dx$$

$$\leq C_p \int (f^*)^p \, M_d(v) \, dx,$$

since $f_{(k,j)}^* \leq 2f^*$ and $\sum_{k,j} 2^k \chi_{I_j^k} \leq 4M_d(v)$.

So let's show 3.29.

If we knew that

$$\int_{I_j^k} (S(f_{(k,j)}))^p \, v \, dx \leq C2^k \int_{I_j^k} (S(f_{(k,j)}))^p \, dx, \qquad (3.30)$$

we'd be done, because we already know that

$$\int_{I_j^k} (S(f_{(k,j)}))^p \, dx \leq C_p \int_{I_j^k} (f_{(k,j)}^*)^p \, dx.$$

Inequality 3.30 isn't hard. Look at $(S(f_{(k,j)}))^p$. It is a function of the form

$$\left(\sum_{I \in \mathcal{F}} \gamma_I \chi_I \right)^\alpha,$$

where $\alpha > 0$, the γ_I's are non-negative, and $\mathcal{F}$ is a bounded family of dyadic intervals such that $v_I \leq C2^k$ for each $I \in \mathcal{F}$. Inequality 3.30 will follow if we know that

$$\int \left(\sum_{I \in \mathcal{F}} \gamma_I \chi_I \right)^\alpha v \, dx \leq C2^k \int \left(\sum_{I \in \mathcal{F}} \gamma_I \chi_I \right)^\alpha dx,$$

and this turns out to be true.

We can phrase this fact in a lemma.

Lemma 3.5. *Let $\mathcal{F}$ be a bounded family of dyadic intervals and let v be a weight such that $v_I \leq A$ for all $I \in \mathcal{F}$, where A is a non-negative number. If $\{\gamma_I\}_{I \in \mathcal{F}}$ is any collection of non-negative numbers and $\alpha > 0$ then*

$$\int \left(\sum_{I \in \mathcal{F}} \gamma_I \chi_I \right)^\alpha v \, dx \leq A \int \left(\sum_{I \in \mathcal{F}} \gamma_I \chi_I \right)^\alpha dx.$$

Proof of Lemma 3.5. There is nothing to prove if $\alpha = 1$. But the general case reduces to $\alpha = 1$ by writing

$$\left(\sum_{I \in \mathcal{F}} \gamma_I \chi_I \right)^\alpha = \sum_{I \in \mathcal{F}} \tilde{\gamma}_I \chi_I,$$

where

$$\tilde{\gamma}_I = \left(\sum_{\substack{J \in \mathcal{F} \\ I \subset J}} \gamma_J \chi_J\right)^\alpha - \left(\sum_{\substack{J \in \mathcal{F} \\ I \subset J, I \neq J}} \gamma_J \chi_J\right)^\alpha.$$

All we have done (let the reader note) is to write the function

$$\left(\sum_{I \in \mathcal{F}} \gamma_I \chi_I\right)^\alpha$$

as a telescoping series. This proves Lemma 3.5 and Theorem 3.11.

What about $p > 2$? The theory we have developed in this chapter implies the following chain of inequalities, valid for all $1 < p < \infty$ and any $r > 1$. The reader is invited to provide the justifications:

$$\int (S(f))^p \, v \, dx \leq \int (S(f))^p \, M_{r,d}(v) \, dx$$

$$\leq C_{p,r} \int (L_a(f))^p \, M_{r,d}(v) \, dx$$

$$\leq C_{p,r} \int (M_d(f))^p \, M_{r,d}(v) \, dx$$

$$\leq C_{p,r} \int |f(x)|^p \, M_d(M_{r,d}(v)) \, dx$$

$$\leq C_{p,r} \int |f(x)|^p \, M_{r,d}(v) \, dx,$$

yielding

$$\int (S(f))^p \, v \, dx \leq C_{p,r} \int |f(x)|^p \, M_{r,d}(v) \, dx$$

for all $p > 1$. If $p \leq 2$, we know that we can replace $M_{r,d}(v)$ with $M_{1,d}(v) = M_d(v)$. It is natural to ask whether we can do so when $p > 2$, and perhaps somewhat unfortunate to find out that we can't.

For $N \geq 0$, let $\tilde{g}_N$ be as we defined it near the end of chapter 2; i.e.,

$$\tilde{g}_N \equiv \sum_0^N (-1)^k \tilde{\lambda}_k h_{(I_k)},$$

where $I_k = [0, 2^k)$ and $\tilde{\lambda}_k \equiv 2^{k/2}/\sqrt{k+1}$. Put $v = \chi_{[0,1)}$. It is easy to see that $M_d(v) = 2^{-k}$ on $[2^{k-1}, 2^k)$ ($k \geq 1$), and therefore that $\int_{2^{k-1}}^{2^k} M_d(v) \, dx = 1/2$. On the other hand, if $x \in [2^{k-1}, 2^k)$ and $1 \leq k \leq N$, then

$$|\tilde{g}_N(x)| = \left|\frac{1}{\sqrt{N+1}} - \frac{1}{\sqrt{N}} + \frac{1}{\sqrt{N-1}} - \frac{1}{\sqrt{N-2}} + \cdots \pm \frac{1}{\sqrt{k+1}}\right| \leq \frac{2}{\sqrt{k+1}}.$$

Therefore

$$\int |\tilde{g}_N|^p \, M_d(v) \, dx \le C \sum_0^N (k+1)^{-p/2} \le C_p,$$

independent of N, because $p > 2$. However, as we saw in chapter 2,

$$\int (S(\tilde{g}_N))^p \, v \, dx \to \infty$$

as $N \to \infty$.

This example shows that a more fully developed, "large p" L^p theory of these inequalities really is necessary. And now would seem to be the right time to start. Alas, such inequalities turn out to be not so simple, and their extension to $p > 2$ will have to wait until we have become acquainted with the theory of Orlicz spaces.

However, before doing that, we will make a quick detour into higher dimensions, and into continuous analogues of the square function.

Exercises

3.1. Let $p(x)$ be any non-trivial polynomial. Show that $v(x) \equiv |p(x)|$ is an A_∞ weight. Show that any such v satisfies 3.12 with a constant A only depending on p's degree. Generalize this result to functions of the form $|p(x)|^\alpha$, where α is a positive number.

3.2. Show that, if v is any weight, then $v \in A_\infty^d$ if and only if there are a constant K and a number $r > 1$ such that

$$\left(\frac{1}{|I|} \int_I (v(x))^r \, dx \right)^{1/r} \le \frac{K}{|I|} \int_I v(x) \, dx \tag{3.31}$$

for all $I \in \mathcal{D}$. In the weighted-norm business, 3.31 is called the "reverse Hölder inequality." (Hint: Assume $|I| = v_I = 1$. The key step is to show that, for some $A > 1$ and some $0 < \eta < 1$,

$$\int_{\{x:\ v(x) > A^k\}} v(x) \, dx \le \eta^k$$

for all $k \ge 0$, because it implies that

$$\sum_0^\infty \int_{A^k < v(x) \le A^{k+1}} (v(x))^{1+\epsilon} \, dx \le \sum_0^\infty A^{(k+1)\epsilon} \eta^k,$$

which converges if ϵ is small enough.) Use the reverse Hölder inequality to prove another characterization of A_∞^d: $v \in A_\infty^d$ if and only if there are positive constants a and b so that, for all dyadic intervals I and all measurable $E \subset I$,

$$v(E) \leq a \left(\frac{|E|}{|I|} \right)^b v(I).$$

(These results are proved in a later chapter.)

3.3. Recall the definition of A_∞ (without the superscript d) from the last chapter. Show that a weight v belongs to A_∞ if and only if there is a constant A such that, for all bounded intervals I,

$$\int_I M(v\chi_I)(x)\, dx \leq A \int_I v(x)\, dx.$$

Then show that $v \in A_\infty$ if and only if there are a constant K and a number $r > 1$ such that

$$\left(\frac{1}{|I|} \int_I (v(x))^r \, dx \right)^{1/r} \leq \frac{K}{|I|} \int_I v(x)\, dx \tag{3.32}$$

for all bounded intervals I. (These results are proved in a later chapter.)

3.4. Show that if $v \in A_\infty$ then v is doubling; i.e., that there is a constant C so that, for all bounded intervals I, $v(2I) \leq Cv(I)$, where I denotes I's concentric double. (This result is proved in a later chapter.)

3.5. Show that there is a constant C so that, for all f, all weights v, and all $\lambda > 0$,

$$v\left(\{x : \ M(f)(x) > \lambda\}\right) \leq \frac{C}{\lambda} \int |f|\, M(v)\, dx,$$

with the consequence that, if $1 < p < \infty$, there is a C_p such that

$$\int (M(f))^p \, v\, dx \leq C_p \int |f(x)|^p \, M(v)\, dx$$

for all functions f and weights v. In other words, show that the non-dyadic analogues of Lemma 3.3 and Corollary 3.6 are also true.

3.6. Here is one way to dispense with the hypothesis 1.4 in Theorem 3.10. For N a positive integer, define the restricted square function of f by the equation

$$S_N(f)(x) \equiv \left(\sum_{I \in \mathcal{D}(N)} \frac{|\lambda_I(f)|^2}{|I|} \chi_I(x) \right)^{1/2}.$$

Notice that $S_N(f) = S_N(f\chi_{I_N})$, where $I_N = [-2^N, 2^N)$, and that $f\chi_{I_N}$ satisfies 1.4. It is now easy to adapt the proof of Theorem 3.10 to show that, for all N and all $1 < p \leq 2$,

$$\int (S_N(f))^p \, v\, dx \leq C(p) \int |f\chi_{I_N}|^p \, M(v)\, dx.$$

Theorem 3.10 (unrestricted form) follows by letting $N \to \infty$. (The reader is invited to recall the appropriate limit theorem.)

3.7. We proved the inequalities,

$$\|f\|_p \sim \|S(f)\|_p, \tag{3.33}$$

valid for $1 < p < \infty$, by means of good-λ inequalities. This approach has the advantage that it generalizes easily to $0 < p \leq 1$ and to weighted problems. But if we only want to know 3.33 for $1 < p < \infty$, we don't have to work so hard. We will sketch the argument, inviting the reader to fill in the details. The inequalities,

$$\int (S(f))^p \, v \, dx \leq C(p) \int |f(x)|^p \, M(v) \, dx,$$

valid for $1 < p \leq 2$, imply that $\|S(f)\|_p \leq C(p)\|f\|_p$ for all $1 < p < \infty$. (Hint: If $2 < p < \infty$,

$$\|S(f)\|_p = \left[\sup\left\{ \int (S(f))^2 \, v \, dx : \|v\|_r \leq 1 \right\} \right]^{1/2},$$

where $r = (p/2)'$, the dual exponent to $p/2$.) If $1 < p < \infty$, f is a finite linear sum of Haar functions,

$$f = \sum \lambda_I(f) h_{(I)},$$

and $\|h\|_{p'} = 1$, where p' is p's dual exponent, then

$$\left| \int f \, h \, dx \right| \leq \sum |\lambda_I(f)| |\lambda_I(h)|$$

$$= \int \left(\sum \frac{|\lambda_I(f)| |\lambda_I(h)|}{|I|} \chi_I \right) dx$$

$$\leq \int S(f) \, S(h) \, dx$$

$$\leq \|S(f)\|_p \|S(h)\|_{p'}$$

$$\leq C(p') \|S(f)\|_p,$$

implying $\|f\|_p \leq C(p')\|S(f)\|_p$ for finite linear sums. How to extend this to general f is left to the reader.

3.8. Show that having $v \in A_\infty$ does not imply

$$\int (S(f))^p \, v \, dx \leq C \int |f(x)|^p \, v \, dx,$$

for *any* $0 < p < \infty$, even if f is bounded and has compact support. (Hint: A good place to look for "bad" A_∞ weights is among the powers $|x|^R$, with R very large, depending on p.)

3.9. We proved Theorem 3.7 (respectively, Theorem 3.8) under the simplifying assumption that $Y(I, v)$ (respectively, $Y_\eta(I, v)$) was finite for all $I \in \mathcal{D}$. Justify those assumptions.

Notes

Theorem 3.2 and inequality 3.3 are from [10]. The Y-functional $Y(I,v)$ first appears in [60]; it is further developed in [62]. Theorem 3.7 is implicit in [62] but is first stated explicitly in [64]. Theorem 3.9 is essentially from [61]. Lemma 3.3 and Theorem 3.6 are both from [18]. The extension of 3.3 to $1 < p < 2$ is based on the main result from [12], which treats a "continuous" version of the square function (we deal with this in a later chapter). The authors there also show that the inequality fails for $p > 2$, but they prove that the result can be salvaged if $M_d(v)$ is replaced by $(M_d(v))^{p/2}v^{1-p/2}$. The corresponding inequality for $0 < p \leq 1$ is proved in [63].

Many Dimensions; Smoothing

If $I \subset \mathbf{R}$ is a bounded interval (not necessarily dyadic), we say that a function $a_{(I)}(x)$ is *adapted* to I if:

   a) supp $a_{(I)} \subset I$;
   b) $\int a_{(I)} \, dx = 0$;
   c) for all x and y in $\mathbf{R}$,

$$|a_{(I)}(x) - a_{(I)}(y)| \leq |I|^{-1/2} \left(\frac{|x - y|}{\ell(I)} \right).$$

Remark. The reader might wonder why we wrote condition c) in this peculiar way; since, for an interval, $|I|$ and $\ell(I)$ are identical. We did so because this definition readily extends to higher dimensions, and we'll be looking at those shortly.

We shall consider how our estimates for finite sums $\sum_{I \in \mathcal{D}} \lambda_I h_{(I)}$ may be extended to finite sums $\sum_{I \in \mathcal{G}} \gamma_I a_{(I)}$, where $\mathcal{G}$ is a family with some especially good properties.

Definition 4.1. *A family of bounded intervals $\mathcal{G}$ is called good if: 1) for every $I \in \mathcal{G}$, I_l and I_r both belong to $\mathcal{G}$; 2) every $I \in \mathcal{G}$ is the right or left half of some other interval in $\mathcal{G}$; 3) if I and J belong to $\mathcal{G}$, then $I \subset J$, $J \subset I$, or $I \cap J = \emptyset$.*

It turns out that a finite linear sum of adapted functions, $\sum_{I \in \mathcal{F}} \gamma_I a_{(I)}$, indexed over a good family $\mathcal{F}$, is almost as easy to work with as a finite linear sum of Haar functions $\sum_{I \in \mathcal{D}} \lambda_I h_{(I)}$. Perhaps the simplest way to see this is by way of $\mathcal{F}$-adapted "Haar functions."

Definition 4.2. *Let $\mathcal{F}$ be a good family. For $I \in \mathcal{F}$, we set*

$$h_{\mathcal{F},(I)}(x) = \begin{cases} |I|^{-1/2} & \textit{if } x \in I_l; \\ -|I|^{-1/2} & \textit{if } x \in I_r; \\ 0 & \textit{otherwise.} \end{cases}$$

Corresponding to these are the $\mathcal{F}$-adapted square function and maximal function.

Definition 4.3. *If* $f = \sum_{I \in \mathcal{F}} \lambda_I h_{\mathcal{F},(I)}$ *is a finite sum, where* $\mathcal{F}$ *is a good family, then*

$$S_{\mathcal{F}}(f)(x) \equiv \left(\sum_{x \in I \in \mathcal{F}} \frac{|\lambda_I|^2}{|I|} \right)^{1/2}.$$

Definition 4.4. *If* $\mathcal{F}$ *is a good family and* $x \in \cup_{\mathcal{F}} I$, *we define*

$$f_{\mathcal{F}}^*(x) \equiv \sup_{I : x \in I \in \mathcal{F}} \frac{1}{|I|} \left| \int_I f \, dt \right|. \tag{4.1}$$

We similarly define

$$M_{\mathcal{F}}(f)(x) \equiv |f|^*(x),$$

the $\mathcal{F}$-*adapted "dyadic" Hardy-Littlewood maximal function.*

(We drop the subscript $\mathcal{F}$ when $\mathcal{F} = \mathcal{D}$. We shall follow this convention in higher dimensions as well: see below.)

The estimates we proved earlier for $S(f)$ and f^*, regarding finite sums $\sum_{I \in \mathcal{D}} \lambda_I h_{(I)}$, hold for $S_{\mathcal{F}}(f)$, $f_{\mathcal{F}}^*$, and finite sums $\sum_{I \in \mathcal{F}} \lambda_I h_{\mathcal{F},(I)}$, with proofs that are identical, except for (what we hope are) obvious modifications. What are "obvious" modifications? Well, instead of having

$$\int (S(f)(x))^2 \, v \, dx \le C \int |f(x)|^2 \, M_d(v) \, dx,$$

we have, for example,

$$\int (S_{\mathcal{F}}(f)(x))^2 \, v \, dx \le C \int |f(x)|^2 \, M_{\mathcal{F}}(v) \, dx,$$

with a constant C that is independent of $\mathcal{F}$. Or, suppose that

$$f = \sum_{I \in \mathcal{G} \subset \mathcal{F}} \lambda_I h_{\mathcal{F},(I)},$$

is a finite sum, and every $I \in \mathcal{G}$ satisfies $Y_\eta(I, v) \le A$. (Recall the definition 3.22 of $Y_\eta(I, v)$.) Then

$$\int (f_{\mathcal{F}}^*(x))^p \, v \, dx \le C_{\eta,p} A^{p/(2\eta)} \int (S_{\mathcal{F}}(f)(x))^p \, v \, dx, \tag{4.2}$$

with $C_{\eta,p}$ only depending on η and p $(0 < p < \infty)$. We shall refer to such generalizations *casually*, in passing. We vigorously encourage the reader to check up on us occasionally to make sure we aren't cheating.

We have need now to speak of functions f that are not expressible as finite sums $\sum_{I \in \mathcal{F}} \lambda_I h_{\mathcal{F},(I)}$ but are still "nice": i.e., controllable by a square function based on the λ_I's. Let's say that f is *nice*, relative to $\mathcal{F}$, if there is an increasing sequence of **finite** families $\mathcal{F}_1 \subset \mathcal{F}_2 \subset \cdots \mathcal{F}$ such that $\mathcal{F} = \cup_k \mathcal{F}_k$ and

$$f = \lim_{k \to \infty} \sum_{I \in \mathcal{F}_k} \langle f, h_{\mathcal{F},(I)} \rangle h_{\mathcal{F},(I)}$$

almost everywhere. For such an f, we extend our square function definition in the obvious way, and set

$$S_{\mathcal{F}}(f)(x) \equiv \left(\sum_{I \in \mathcal{F}} \frac{|\langle f, h_{\mathcal{F},(I)} \rangle|^2}{|I|} \chi_I(x) \right)^{1/2}.$$

By Fatou's Lemma, any inequality of the form

$$\int (f^*(x))^p \, v \, dx \leq C \int (S_{\mathcal{F}}(f)(x))^p \, w \, dx,$$

valid for finite sums $f = \sum_{I \in \mathcal{F}} \lambda_I h_{\mathcal{F},(I)}$ $(= \sum_{I \in \mathcal{F}} \langle f, h_{\mathcal{F},(I)} \rangle h_{\mathcal{F},(I)})$, holds automatically for "nice" f's, with the same constant C.

The "nice" functions we have in mind are finite linear sums of adapted functions, $f = \sum_{I \in \mathcal{F}} \gamma_I a_{(I)}$, indexed over a good family $\mathcal{F}$. To see that these are "nice", consider a single function $a_{(I)}$, where $I \in \mathcal{F}$. We leave it as an exercise to show that

$$a_{(I)}(x) = \lim_{k \to \infty} \sum_{\substack{J \in \mathcal{F}: J \subset 2^k I \\ \ell(J) \geq 2^{-k}}} \langle a_{(I)}, h_{\mathcal{F},(J)} \rangle h_{\mathcal{F},(J)} \qquad (4.3)$$

almost everywhere (and indeed, in L^∞). (The reader will have an easier time proving this if he first observes that $\langle a_{(I)}, h_{\mathcal{F},(J)} \rangle = 0$ for $J \not\subset I$.) The families $\{ J \in \mathcal{F} : J \subset 2^k I, \, \ell(J) \geq 2^{-k} \}$ are clearly finite, increasing, and have union equal to $\mathcal{F}$.

For such a function $f = \sum_{I \in \mathcal{F}} \gamma_I a_{(I)}$, we define a "semi-discrete" square function by:

$$S_{sd,\mathcal{F}}(f)(x) \equiv \left(\sum_{I \in \mathcal{F}} \frac{|\gamma_I|^2}{|I|} \chi_I(x) \right)^{1/2}. \qquad (4.4)$$

Our first order of business is to establish an inequality between $S_{sd,\mathcal{F}}(f)$ and $S_{\mathcal{F}}(f)$.

Theorem 4.1. *There is an absolute constant C so that, for all good families $\mathcal{F}$ and all finite sums $f = \sum_{I \in \mathcal{F}} \gamma_I a_{(I)}$,*

$$S_{\mathcal{F}}(f)(x) \leq C \, S_{sd,\mathcal{F}}(f)(x)$$

pointwise.

Proof of Theorem 4.1. Temporarily fix $J \in \mathcal{F}$. We must estimate $\lambda_J \equiv \langle f, h_{\mathcal{F},(J)} \rangle$. Since—as noted above—$\langle a_{(I)}, h_{\mathcal{F},(J)} \rangle = 0$ for $J \not\subset I$, we may write

$$|\lambda_J| \leq \sum_{I: J \subset I} |\gamma_I| |\langle a_{(I)}, h_{\mathcal{F},(J)} \rangle|.$$

Temporarily fix an I containing J. Because $\int h_{\mathcal{F},(J)} \, dx = 0$,

$$|\langle a_{(I)}, h_{\mathcal{F},(J)} \rangle| = \left| \int_J a_{(I)}(x) \, h_{\mathcal{F},(J)}(x) \, dx \right|$$

$$= \left| \int_J (a_{(I)}(x) - a_{(I)}(x_J^*)) \, h_{\mathcal{F},(J)}(x) \, dx \right| \qquad (4.5)$$

where x_J^* is some fixed (but arbitrary) point in J. Our estimate of $a_{(I)}$'s smoothness (condition c)) says that

$$|a_{(I)}(x) - a_{(I)}(x_J^*)| \leq |I|^{-1/2} \left(\frac{|x - x_J^*|}{\ell(I)} \right) \leq |I|^{-1/2} (\ell(J)/\ell(I))$$

for all $x \in J$. Therefore,

$$\left| \int_J (a_{(I)}(x) - a_{(I)}(x_J^*)) \, h_{\mathcal{F},(J)}(x) \, dx \right| \leq (\ell(J)/\ell(I)) |I|^{-1/2} \|h_{\mathcal{F},(J)}\|_1$$

$$= (\ell(J)/\ell(I))(\sqrt{|J|}/\sqrt{|I|}).$$

Plugging this in and summing, we get

$$\frac{|\lambda_J|^2}{|J|} \leq \left(\sum_{I: J \subset I} \frac{|\gamma_I|}{\sqrt{|I|}} (\ell(J)/\ell(I)) \right)^2,$$

which is less than or equal to

$$\left(\sum_{I: J \subset I} \frac{|\gamma_I|^2}{|I|} (\ell(J)/\ell(I)) \right) \left(\sum_{I: J \subset I} (\ell(J)/\ell(I)) \right). \qquad (4.6)$$

But, for each $k = 0, 1, 2, \ldots$, there is exactly one $I \in \mathcal{F}$ such that $J \subset I$ and $\ell(I) = 2^k \ell(J)$. Therefore $\sum_{I: J \subset I} (\ell(J)/\ell(J)) \leq 2$, and we can bound 4.6 by

$$2 \sum_{I: J \subset I} \frac{|\gamma_I|^2}{|I|} (\ell(J)/\ell(I)).$$

Now we sum over all J's containing a fixed $x \in \mathbf{R}$—and take care to remember that if $x \in J \subset I$, then $x \in I$:

$$(S_{\mathcal{F}}(f)(x))^2 = \sum_{J:x\in J} \frac{|\lambda_J|^2}{|J|}$$

$$\leq 2 \sum_{J:x\in J} \left(\sum_{I:J\subset I} \frac{|\gamma_I|^2}{|I|}(\ell(J)/\ell(I)) \right)$$

$$= 2 \sum_{I:x\in I} \frac{|\gamma_I|^2}{|I|} \left(\sum_{J:x\in J\subset I} (\ell(J)/\ell(I)) \right)$$

$$\leq 4 \sum_{I:x\in I} \frac{|\gamma_I|^2}{|I|}$$

$$= 4(S_{sd,\mathcal{F}}(f)(x))^2,$$

where the next-to-last inequality comes from the fact that

$$\sum_{J:x\in J\subset I} (\ell(J)/\ell(I)) \leq 2;$$

because, for each $k = 0, 1, 2, \ldots$, there is only one $J \subset I$ that contains x and satisfies $\ell(J) = 2^{-k}\ell(I)$. This proves Theorem 4.1.

Before continuing, we want to take special notice of equation 4.5, which is the key to Theorem 4.1. This is actually a particular example of a general technique, in which a cancelation condition in one function ($h_{\mathcal{F},(J)}$ in this case) is played off against smoothness in another, such as $a_{(I)}$. We shall enshrine this idea in a lemma.

Lemma 4.1. *Let f and g be locally integrable functions defined on a cube $Q \subset \mathbf{R}^d$. Suppose that g's support is contained in Q, $\int_Q g\,dx = 0$, and, for some positive numbers A and α,*

$$|f(x) - f(y)| \leq A|x-y|^\alpha$$

for all x and y in Q. Then:

$$\left| \int f(x)\,g(x)\,dx \right| \leq C_{d,\alpha} A\ell(Q)^\alpha \int_Q |g(x)|\,dx,$$

where we can take $C_{d,\alpha}$ to be $((1/2)\sqrt{d})^\alpha$.

Proof. Let x_Q be the center of Q. Because of g's cancelation,

$$\left| \int f(x)\,g(x)\,dx \right| = \left| \int (f(x) - f(x_Q))\,g(x)\,dx \right|.$$

However, for all $x \in Q$,

$$|f(x) - f(x_Q)| \leq A|x - x_Q|^\alpha \leq A((1/2)\sqrt{d})^\alpha \ell(Q)^\alpha,$$

from which the inequality follows.

Remark. The point of Lemma 4.1 is not the value of the constant $C_{d,\alpha}$, but the factor $\ell(Q)^\alpha$, which goes to 0 as $\ell(Q) \to 0$. Had we chosen a different fixed point in Q (as we did in the proof of Theorem 4.1), we would have got a slightly bigger constant $C_{d,\alpha}$, but the same factor of $\ell(Q)^\alpha$.

Theorem 4.1 implies easy corollaries of our main results from the preceding chapter. For example, we obtain:

Theorem 4.2. *Let $0 < p < \infty$ and let $\tau > p/2$. Let $\mathcal{F}$ be a good family and let v and w be weights such that*

$$\int_I v(x) \left(\log(e + v(x)/v_I)\right)^\tau dx \le \int_I w(x)\, dx$$

for all $I \in \mathcal{F}$. Then, for all finite sums $f = \sum_{I \in \mathcal{F}} \gamma_I a_{(I)}$, where each $a_{(I)}$ is adapted to I,

$$\int (f_{\mathcal{F}}^*)^p\, v\, dx \le C \int (S_{sd,\mathcal{F}}(f))^p\, w\, dx,$$

where the constant C only depends on p and τ.

Also:

Corollary 4.1. *Let $f = \sum_{I \in \mathcal{F}} \gamma_I a_{(I)}$ be as in the statement of Theorem 4.2, let v be a weight, and let $0 < p < \infty$. If $0 < p < 2$ then*

$$\int (f_{\mathcal{F}}^*)^p\, v\, dx \le C(p) \int (S_{sd,\mathcal{F}}(f))^p\, M_{\mathcal{F}}(v)\, dx.$$

Assuming that we can prove the appropriate Orlicz space results, we will also have this generalization of Corollary 3.5:

Corollary 4.2. *Let $0 < p < \infty$. There is a positive constant $C = C(p)$ such that, for all weights v and all finite sums $f = \sum_{I \in \mathcal{F}} \gamma_I a_{(I)}$,*

$$\int (f_{\mathcal{F}}^*)^p\, v\, dx \le C \int (S_{sd,\mathcal{F}}(f))^p\, M_{\mathcal{F}}^{[p/2]+1}(v)\, dx.$$

For many purposes, these generalizations are entirely adequate. However, when we come to study weighted inequalities of the form

$$\int |f|^2\, v\, dx \le \int |\nabla f|^2\, dx,$$

we will want a little more control than this straightforward approach provides. That is why we will prove the following, more delicate estimate.

Theorem 4.3. *Let $\mathcal{F}$ be a good family, and suppose that $\mathcal{G} \subset \mathcal{F}$ is such that $Y_\eta(I, v) \le A$ for all $I \in \mathcal{G}$. Let $f = \sum_{I \in \mathcal{G}} \gamma_I a_{(I)}$ be a finite sum. Then*

$$\int |f|^p\, v\, dx \le C A^{p/(2\eta)} \int (S_{sd,\mathcal{F}}(f))^p\, v\, dx,$$

where C depends on p and η, and $0 < p < \infty$.

Proof of Theorem 4.3. We control $f_{\mathcal{F}}^*(x)$ via a good-λ inequality. There is a small subtlety here. We cannot simply use inequality 4.2 and Theorem 4.1 to get

$$\int (f_{\mathcal{F}}^*)^p \, v \, dx \leq CA^{p/(2\eta)} \int (S_{\mathcal{F}}(f))^p \, v \, dx$$

$$\leq CA^{p/(2\eta)} \int (S_{sd,\mathcal{F}}(f))^p \, v \, dx,$$

because, if we write $f = \sum_{I \in \mathcal{F}} \lambda_I h_{\mathcal{F},(I)}$, it's almost certain that some λ_Is will be non-zero for Is that don't belong to $\mathcal{G}$. Therefore we have to directly control the partial sums of $\sum_{I \in \mathcal{G}} \gamma_I a_{(I)}$. We can't assume that these will be constant across our stopping intervals I_i^{λ}, but the smoothness of the $a_{(I)}$s implies that they will be almost constant, and that's good enough. The crucial observation is the following. Let $J \in \mathcal{F}$ and set

$$g = \sum_{\substack{I \in \mathcal{G} \\ J \subset I}} \gamma_I a_{(I)}.$$

Define

$$R = \left(\sum_{\substack{I \in \mathcal{G} \\ J \subset I}} \frac{|\gamma_I|^2}{|I|} \right)^{1/2}.$$

Then, for any x and y in J,

$$|g(x) - g(y)| \leq CR. \tag{4.7}$$

The proof of this inequality is very simple. For any x and y in J, we have

$$|g(x) - g(y)| \leq \sum_{\substack{I \in \mathcal{G} \\ J \subset I}} |\gamma_I| |a_{(I)}(x) - a_{(I)}(y)|$$

$$\leq \sum_{\substack{I \in \mathcal{G} \\ J \subset I}} |\gamma_I| |I|^{-1/2} (|x - y|/\ell(I))$$

$$\leq \ell(J) \sum_{\substack{I \in \mathcal{G} \\ J \subset I}} |\gamma_I| |I|^{-1/2} \ell(I)^{-1}$$

$$\leq R \left(\sum_{\substack{I \in \mathcal{G} \\ J \subset I}} (\ell(J)/\ell(I))^2 \right)^{1/2},$$

where the last inequality is due to Cauchy-Schwarz. Now 4.7 follows from the fact that

$$\sum_{\substack{I \in \mathcal{G} \\ J \subset I}} (\ell(J)/\ell(I))^2 \leq 1 + \frac{1}{4} + \frac{1}{16} + \cdots = 4/3.$$

With 4.7 in hand, let us now consider a maximal $I_i^\lambda \in \mathcal{F}$ such that $|f_{I_i^\lambda}| > \lambda$. Let P be I_i^λ's immediate (relative to $\mathcal{F}$) superinterval. Then $|f_P| \leq \lambda$ and $||f_P| - |f_{I_i^\lambda}|| \leq \sup_{x,y \in P} |\tilde{g}(x) - \tilde{g}(y)|$, where

$$\tilde{g}(x) = \sum_{\substack{I \in \mathcal{G} \\ I_i^\lambda \subset I}} \gamma_I a_{(I)}.$$

Because of 4.7, if $|f_{I_i^\lambda}| > 1.1\lambda$, then we have (setting $J = I_i^\lambda$) $R > c\lambda$, implying that $S_{sd,\mathcal{F}}(f) > c\lambda$ on all of I_i^λ, and we have nothing to prove. For the same reason, inequality 4.7 lets us assume that

$$\left| \sum_{\substack{I \in \mathcal{F} \\ I_i^\lambda \subset I}} \gamma_I a_{(I)} \right| \leq 1.1\lambda$$

on all of I_i^λ. Thus, we only need to worry about the summands which are supported in I_i^λ. Set

$$h = \sum_{\substack{I \in \mathcal{F} \\ I \subset I_i^\lambda}} \gamma_I a_{(I)}.$$

Following our earlier procedure (in the "pure dyadic" case), we need to show that

$$v\left(\{ x \in I_i^\lambda : h_{\mathcal{F}}^*(x) > .9\lambda, \ S_{sd,\mathcal{F}}(f) \leq \gamma\lambda \} \right) \leq \epsilon(p) v(I_i^\lambda)$$

for $\gamma \sim A^{-1/(2\eta)}$. As in the dyadic case, the problem is that I_i^λ probably doesn't satisfy $Y_\eta(I_i^\lambda, v) \leq A$. But, just as in the dyadic case, the solution presents itself. Let $\{J_k\}_k$ be the maximal $J \in \mathcal{G}$ which are also contained in I_i^λ. Because of the cancelation in the $a_{(I)}$'s, if $K \in \mathcal{F}$ but K is not properly contained in any J_k, we must have $\int_K h = 0$. Therefore,

$$v\left(\{ x \in I_i^\lambda : h_{\mathcal{F}}^*(x) > .9\lambda, \ S_{sd,\mathcal{F}}(f) \leq \gamma\lambda \} \right)$$
$$= \sum_k v\left(\{ x \in J_k : h_{\mathcal{F}}^*(x) > .9\lambda, \ S_{sd,\mathcal{F}}(f) \leq \gamma\lambda \} \right),$$

but we *do* have our estimate $Y_\eta(J_k, v) \leq A$ for each k. An application of Theorem 4.1, combined with the dyadic exponential square good-λ inequality, yields

$$v\left(\{ x \in J_k : h_{\mathcal{F}}^*(x) > .9\lambda, \ S_{sd,\mathcal{F}}(f) \leq \gamma\lambda \} \right) \leq \epsilon(p) v(J_k)$$

for each k, for $\gamma \sim A^{-1/(2\eta)}$. Summing on k now yields our inequality, and finishes the proof of Theorem 4.3.

By following the pattern of Theorem 3.7, we obtain:

Corollary 4.3. *Let $\tau > 1$. There is a constant $C = C(\tau)$ such that, for all weights v and all finite linear sums $f = \sum_{I \in \mathcal{F}} \lambda_I a_{(I)}$, where the $a_{(I)}$'s are adapted to intervals I belong to a good family $\mathcal{F}$,*

$$\int |f(x)|^2 \, v \, dx \le C \sum_{I \in \mathcal{F}} \frac{|\lambda_I|^2}{|I|} \int_I v(x) \, (\log(e + v(x)/v_I))^\tau \, dx.$$

(See also Theorem 4.6 at the end of this chapter.)

Everything we've said so far generalizes with almost no extra work to $\mathbf{R}^d$, $d > 1$. We must replace dyadic intervals with dyadic cubes,

$$Q = [\frac{j_1}{2^k}, \frac{j_1 + 1}{2^k}) \times [\frac{j_2}{2^k}, \frac{j_2 + 1}{2^k}) \times \cdots \times [\frac{j_d}{2^k}, \frac{j_d + 1}{2^k}),$$

where k and the j_i are integers. The sidelength of such a cube, denoted $\ell(Q)$, is 2^{-k}. The cube Q is contained in a unique dyadic cube Q' with sidelength $2^{-(k-1)}$ and it contains 2^d congruent dyadic subcubes of sidelengths $2^{-(k+1)}$. We denote the family of all dyadic cubes by $\mathcal{D}_d$.

There are many (essentially equivalent) ways we can define the dyadic square function for $\mathbf{R}^d$. We elect the following. If $f \in L^1_{\text{loc}}(\mathbf{R}^d)$ and k is an integer, we set

$$f_k = \sum_{Q \in \mathcal{D}_d : \ell(Q) = 2^{-k}} f_Q \chi_Q; \qquad (4.8)$$

i.e., f_k is what we get when we replace f by its averages over dyadic cubes of sidelength 2^{-k}. Following our convention in the 1-dimensional case, we define $f^*(x) \equiv \sup_k |f_k(x)|$. If $Q \in \mathcal{D}_d$ and $\ell(Q) = 2^{-k}$, we set

$$a_{(Q)}(f) = (f_{k+1} - f_k)\chi_Q.$$

The function $a_{(Q)}(f)$: a) is supported in Q; b) is constant on Q's immediate dyadic subcubes; and c) satisfies $\int a_{(Q)}(f) = 0$. Let $\mathcal{H}(Q)$ denote the space of functions satisfying a), b), and c). It has dimension $2^d - 1$. At the end of this chapter we will show how to construct a canonical orthonormal basis $\{h_{(Q),i}\}_1^{2^d-1}$ for it, one that reduces to a single Haar function when $d = 1$. For now we will take the existence of this basis for granted, and note some of its properties:

$$\|h_{(Q),i}\|_\infty \sim |Q|^{-1/2};$$

$$a_{(Q)}(f) = \sum_i \langle f, h_{(Q),i} \rangle h_{(Q),i};$$

$$\|a_{(Q)}(f)\|_2^2 = \sum_i |\langle f, h_{(Q),i} \rangle|^2;$$

$$\|a_{(Q)}(f)\|_2 \sim \|a_{(Q)}(f)\|_\infty |Q|^{1/2}$$
$$\sim \max_i |\langle f, h_{(Q),i} \rangle|.$$

The family $\{a_{(Q)}(f)\}_{Q \in \mathcal{D}_d}$ is pairwise orthogonal. If $f \in L^2(\mathbf{R}^d)$ then $f = \sum_Q a_{(Q)}(f)$ in L^2, implying

$$\int |f|^2 \, dx = \sum_Q \|a_{(Q)}(f)\|_2^2.$$

We rewrite the sum on the right as

$$\int \left(\sum_Q \frac{\|a_{(Q)}(f)\|_2^2}{|Q|} \chi_Q \right) dx.$$

We define our *d-dimensional dyadic square function* by:

$$S_d(f)(x) \equiv \left(\sum_Q \frac{\|a_{(Q)}(f)\|_2^2}{|Q|} \chi_Q(x) \right)^{1/2}. \qquad (4.9)$$

We can make this look more like the 1-dimensional version $S(f)$ (see 2.8) if we write it as:

$$\left(\sum_Q \frac{1}{|Q|} \left(\sum_i |\langle f, h_{(Q),i} \rangle|^2 \right) \chi_Q(x) \right)^{1/2}.$$

What of our exponential-square and weighted norm estimates? For one class—those in which f controls $S_d(f)$—we need only minor changes in the proofs to extend them to d dimensions. The basic weighted L^2 inequality,

$$\int_{\mathbf{R}^d} (S_d(f))^2 \, v \, dx \leq C_d \int_{\mathbf{R}^d} |f|^2 \, M_d(v) \, dx,$$

is proved essentially as in $\mathbf{R}^1$, as are its extensions to L^p ($1 < p < 2$),

$$\int_{\mathbf{R}^d} (S_d(f))^p \, v \, dx \leq C_{d,p} \int_{\mathbf{R}^d} |f|^p \, M_d(v) \, dx$$

and

$$\int_{\mathbf{R}^d} (S_d(f))^p \, v \, dx \leq C_{d,p} \int_{\mathbf{R}^d} (f^*)^p \, M_d(v) \, dx$$

for $0 < p \leq 1$. Following our convention, noted above, we are using f^* to denote the "ordinary" dyadic maximal function of f with respect to $\mathcal{D}_d$:

$$f^*(x) \equiv \sup_{x \in Q \in \mathcal{D}_d} |f_Q|.$$

(We strongly encourage the reader to check that what we have just said is true.)

For the other class, we need to modify the arguments a little more. The heart of the 1-dimensional proof was the observation that if $\int_0^1 f \, dx = 0$ and f is constant on the two halves of $[0,1)$ (say $f(x) \equiv t$ on $[0,1/2)$ and $f(x) \equiv -t$ on $[1/2,1)$), then

$$\int_0^1 \exp(f(x)) \, dx = \cosh(t);$$

and what was really needed was that the integral was $\leq \cosh(t)$. As a substitute for this simple fact we have:

Lemma 4.2. *Let (Ω, μ) be a probability space and let $f : \Omega \mapsto \mathbf{R}$ be a random variable (i.e., a measurable function) such that $|f(x)| \leq t$ for all $x \in \Omega$ and*

$$\int_{\Omega} f(x) \, d\mu(x) = 0.$$

Then

$$\int_{\Omega} \exp(f(x)) \, d\mu(x) \leq \cosh(t).$$

Proof of Lemma 4.2. Let ν be f's distribution measure on $\mathbf{R}$. This means that ν is a Borel probability measure, supported in $[-t, t]$ and satisfying

$$\nu(E) = \mu(\{x \in \Omega : \ f(x) \in E\})$$

for all Borel $E \subset \mathbf{R}$. We recall a theorem from basic probability theory: *If $F : [-t, t] \mapsto \mathbf{R}$ is any continuous function, then*

$$\int_{\Omega} F(f(x)) \, d\mu(x) = \int_{-t}^{t} F(s) \, d\nu(s). \tag{4.10}$$

If the reader hasn't seen 4.10 before, an easy way to prove it is to observe that, for all $-\infty < a < b < \infty$,

$$\mu(\{x \in \Omega : \ a < F(f(x)) < b\}) = \nu(\{s \in [-t, t] : \ a < F(s) < b\}),$$

from which 4.10 readily follows by an approximation argument.

Given 4.10, our hypotheses on f imply that

$$\int_{-t}^{t} x \, d\nu(x) = 0, \tag{4.11}$$

and the conclusion of the lemma is equivalent to

$$\int_{-t}^{t} \exp(x) \, d\nu(x) \leq \cosh(t).$$

We prove the conclusion this way. The function $\exp(x)$ is convex. On $[-t, t]$ it lies below the line connecting $(-t, \exp(-t))$ and $(t, \exp(t))$. This line has equation $y = mx + b$, where b is exactly $\cosh(t)$. Thus:

$$\int_{-t}^{t} \exp(x) \, d\nu(x) \leq \int_{-t}^{t} (mx + b) \, d\nu(x)$$

$$= \int_{-t}^{t} b \, d\nu(x)$$

$$= b = \cosh(t),$$

where the next-to-last line follows from 4.11. That proves Lemma 4.2.

Remark. I learned this proof from Juan Sueiro Bal while he was a graduate student at the University of Wisconsin in Madison.

Now consider one of our difference functions $a_{(Q)}(f)$. It is supported in Q, has integral 0, and satisfies

$$\|a_{(Q)}(f)\|_\infty \leq C(d)\|a_{(Q)}(f)\|_2|Q|^{-1/2}.$$

By Lemma 4.2,

$$\frac{1}{|Q|}\int_Q \exp(a_{(Q)}(f))\,dx \leq \cosh(C(d)\|a_{(Q)}(f)\|_2|Q|^{-1/2}),$$

which is less than or equal to

$$\exp\left(C'(d)\frac{\|a_{(Q)}(f)\|_2^2}{|Q|}\right).$$

The expression

$$\frac{\|a_{(Q)}(f)\|_2^2}{|Q|}$$

corresponds to what we got in the 1-dimensional case:

$$\frac{|\lambda_I|^2}{|I|}.$$

If run our 1-dimensional proof again, but now in d dimensions, we obtain:

Theorem 4.4. *Let Q be a dyadic cube in $\mathbf{R}^d$. Let $f \in L^1(\mathbf{R}^d)$ have support contained in Q and satisfy $\int f\,dx = 0$. Define f_k as in 4.8 and $S_d(f)$ as in 4.9. For all integers k and positive numbers t,*

$$\frac{1}{|Q|}\int_Q \exp(tf_k(x) - C'(d)t^2(S_d(f_k))^2)\,dx \leq 1,$$

where $C'(d)$ only depends on d. There exist positive constants $c_1(d)$ and $c_2(d)$ so that, if $\|S_d(f)\|_\infty \leq 1$ then, for all $\lambda > 0$,

$$|\{x \in Q : |f(x)| > \lambda\}| \leq c_1(d)|Q|\exp(-c_2(d)\lambda^2).$$

The Y-functional and its variants are defined exactly as in $\mathbf{R}^1$. For $\eta > 0$,

$$Y_\eta(Q,v) \equiv \frac{1}{v(Q)}\int_Q v(x)\,(\log(e + v(x)/v_Q))^\eta\,dx \tag{4.12}$$

when $v(Q) > 0$, and is 1 otherwise. Also as in the 1-dimensional case, we set $Y(Q,v) \equiv Y_1(Q,v)$. The proofs from the 1-dimensional setting go through almost verbatim, and yield:

Theorem 4.5. *Let $0 < p < \infty$ and let $\eta > p/2$. If v and w are non-negative functions in $L^1_{\text{loc}}(\mathbf{R}^d)$ that satisfy*

$$\int_Q v(x)\,(\log(e + v(x)/v_Q))^\eta\,dx \le \int_Q w(x)\,dx$$

for all $Q \in \mathcal{D}_d$, then, for all finite linear sums

$$f = \sum_Q a_{(Q)}(f),$$

we have

$$\int_{\mathbf{R}^d} (f^*)^p\,v\,dx \le C(p,\eta,d) \int_{\mathbf{R}^d} (S_d(f))^p\,w\,dx,$$

where $S_d(f)$ is as defined in 4.9.

We continue our generalizing process by defining d-dimensional adapted functions. (We refer the reader to the definition of 1-dimensional adapted functions given at the beginning of the chapter.)

If $Q \subset \mathbf{R}^d$ is a cube (not necessarily dyadic), we say that $b_{(Q)}$ is *adapted* to Q if:

a) supp $b_{(Q)} \subset Q$;

b) $\int b_{(Q)}\,dx = 0$;

c) for all x and y in $\mathbf{R}$,

$$|b_{(Q)}(x) - b_{(Q)}(y)| \le |Q|^{-1/2}\left(\frac{|x-y|}{\ell(Q)}\right).$$

Similarly, we define d-dimensional "good" families of cubes, following the pattern for $\mathbf{R}^1$. The only twist is that, no longer having a right and left half of a cube Q, we speak of the 2^d congruent subcubes obtained by bisecting Q's component intervals. We will refer to these as Q's "immediate dyadic subcubes" even if Q isn't dyadic.

A family of cubes $\mathcal{G}$ is called *good* if: 1) for every $Q \in \mathcal{G}$, all of its immediate dyadic subcubes belong to $\mathcal{G}$; 2) every $Q \in \mathcal{G}$ is an immediate dyadic subcube of some other cube in $\mathcal{G}$; 3) if Q and Q' belong to $\mathcal{G}$, then $Q \subset Q'$, $Q' \subset Q$, or $Q \cap Q' = \emptyset$.

Just as in $\mathbf{R}^1$, we can define $\mathcal{G}$-based versions of our Haar functions, the dyadic maximal function, and the dyadic square function. We will denote these respectively as $h_{\mathcal{G},(Q),i}$, $f^*_{\mathcal{G}}$, and $S_{\mathcal{G},d}$; their definitions should give the reader no problems.

If $\mathcal{G}$ is a good family of cubes, and $f = \sum_{Q \in \mathcal{G}} \lambda_Q b_{(Q)}$ is a finite linear sum of adapted functions, we define the $\mathcal{G}$-based semi-discrete square function by

$$S_{sd,\mathcal{G}}(f)(x) \equiv \left(\sum_{Q \in \mathcal{G}} \frac{|\lambda_Q|^2}{|Q|}\chi_Q(x)\right)^{1/2}.$$

The reader will recall that in one dimension we had $S_{\mathcal{F}}(f)(x) \leq C\, S_{sd,\mathcal{F}}(f)(x)$ pointwise (Theorem 4.1), valid for finite linear sums of $\mathcal{F}$-based adapted functions, and that this was a useful inequality. The corresponding inequality, $S_{\mathcal{G},d}(f)(x) \leq C S_{sd,\mathcal{G}}(f)(x)$, holds in d dimensions, with almost the same proof. The key ingredient is an estimate on $|\langle b_{(Q)}, h_{\mathcal{G},(Q'),i} \rangle|$, where Q and Q' belong to $\mathcal{G}$ and $b_{(Q)}$ is adapted to Q. This estimate is:

$$|\langle b_{(Q)}, h_{\mathcal{G},(Q'),i} \rangle| \begin{cases} = 0 & \text{if } Q' \not\subset Q; \\ \leq C \dfrac{|Q'|^{1/2}}{|Q|^{1/2}} \dfrac{\ell(Q')}{\ell(Q)} & \text{if } Q' \subset Q, \end{cases}$$

which has a close analogue (and almost identical proof) to one obtained in the proof of Theorem 4.1. Working out this estimate and how it implies that $S_{\mathcal{G},d}(f)(x) \leq C S_{sd,\mathcal{G}}(f)(x)$ will make a good exercise for the reader. He should also have no trouble proving the following theorems.

Theorem 4.6. *If $\tau > 1$, there is a constant $C = C(\tau,d)$ such that, for all weights v, all good families $\mathcal{G}$, and all finite linear sums $f = \sum_{Q \in \mathcal{G}} \lambda_Q b_{(Q)}$, where each $b_{(Q)}$ is adapted to Q,*

$$\int_{\mathbf{R}^d} |f(x)|^2\, v\, dx \leq C \sum_{Q \in \mathcal{G}} \frac{|\lambda_Q|^2}{|Q|} \int_Q v(x)\, (\log(e + v(x)/v_Q))^\tau\, dx.$$

Theorem 4.7. *If $0 < p < \infty$, $\eta > p/2$, and v and w are weights such that*

$$\int_Q v(x)\, (\log(e + v(x)/v_Q))^\eta\, dx \leq \int_Q w(x)\, dx$$

for all cubes $Q \in \mathcal{G}$, a good family, then, for all finite linear sums $f = \sum_{Q \in \mathcal{G}} \lambda_Q b_{(Q)}$, where each $b_{(Q)}$ is adapted to Q, we have

$$\int_{\mathbf{R}^d} (f_{\mathcal{G}}^*)^p\, v\, dx \leq C(p,\eta,d) \int_{\mathbf{R}^d} \tilde{S}_{sd,\mathcal{G}}^p(f)\, w\, dx.$$

We will use Theorem 4.6 when we look at the Calderón reproducing formula and Schrödinger operators.

Multidimensional "Haar Functions"

Let $Q_0 = [0,1)^d$ be the unit dyadic cube, and consider the space $\mathcal{H}(Q_0)$ of all functions f supported in Q_0 that are constant on Q_0's immediate dyadic subcubes, and which satisfy $\int f\, dx = 0$. This space has an L^2 orthonormal basis of $2^d - 1$ elements. We shall construct a *canonical* basis for $\mathcal{H}(Q_0)$, one that reduces to a single Haar function when $d = 1$. By translating and scaling, we get a corresponding set of "Haar functions" for every cube $Q \subset \mathbf{R}^d$.

The construction depends on the following lemma.

Lemma 4.3. *There exist $2^d - 1$ pairs of subsets of Q_0, (E_j, F_j) ($1 \leq j \leq 2^d - 1$), such that:*

a) each E_j or F_j consists of a union of immediate dyadic subcubes of Q_0;

b) for each j, $E_j \cap F_j = \emptyset$;

c) for each j, $|E_j| = |F_j| > 0$;

d) if $j \neq k$, then either $(E_j \cup F_j) \cap (E_k \cup F_k) = \emptyset$, or $E_j \cup F_j$ is entirely contained in one of E_k or F_k, or $E_k \cup F_k$ is entirely contained in one of E_j or F_j.

Remark. What on earth does this mean? Set

$$h_j(x) = \frac{\chi_{E_j}(x) - \chi_{F_j}(x)}{|E_j \cup F_j|^{1/2}}.$$

Each h_j belongs to $\mathcal{H}(Q_0)$, because of a) and c). Because of b), $\|h_j\|_2 = 1$. If $k \neq j$, then either h_j and h_k have disjoint supports or else one of them—let's say it's h_j—has its support entirely contained in a set across which the other function, h_k, is constant. But $\int h_j \, dx = 0$, implying $\int h_j(x) \, h_k(x) \, dx = 0$. Therefore the family $\{h_j\}_1^{2^d - 1}$ is an orthonormal basis for $\mathcal{H}(Q_0)$—assuming we can prove Lemma 4.3.

Proof of Lemma 4.3. We do induction on d. The result is trivial when $d = 1$. Assume the result for $d - 1$, and let (E_j', F_j') ($1 \leq j \leq 2^{d-1} - 1$) be the corresponding pairs of subsets of $Q_0' \equiv [0, 1)^{d-1}$. Write $Q_0 = Q_0' \times [0, 1)$. For $i = 1, 2$, define

$$E_j^1 = E_j' \times [0, 1/2)$$
$$F_j^1 = F_j' \times [0, 1/2)$$
$$E_j^2 = E_j' \times [1/2, 1)$$
$$F_j^2 = F_j' \times [1/2, 1),$$

and set $G_1 = Q_0' \times [0, 1/2)$ and $G_2 = Q_0' \times [1/2/1)$. Then the pairs (E_j^i, F_j^i) ($i = 1, 2, 1 \leq j \leq 2^{d-1} - 1$) and (G_1, G_2) make up the collection we seek.

Exercises

4.1. Prove equation 4.3.

4.2. Recall Theorem 3.10 and Theorem 3.11. State and prove appropriate d-dimensional generalizations of them.

4.3. State and prove appropriate d-dimensional generalizations of all of the exercises from the preceding chapter.

Notes

The extension of Theorem 3.2 to d dimensions is found in [10]. The idea of treating adapted functions like "nearly" Haar functions has been used by many authors, notably in [9] and [57]. Most of the d-dimensional weighted norm inequalities in this chapter are from [62]. The construction of multi-dimensional Haar functions is from [65].

The Calderón Reproducing Formula I

The theorems at the end of the last chapter dealt with functions f expressible as nice sums of adapted functions. That is, given a function $f = \sum_Q \lambda_Q a_{(Q)}$, where the cubes Q belonged to a good family, we could say something about its size, in various weighted spaces, in terms of the coefficients λ_Q. But those theorems didn't tell us what to do if we were just given the function f.

In essence, the Calderón reproducing formula allows us to express "arbitrary" functions as sums of adapted functions, indexed over good families of cubes, and with useful (or at least intelligible) bounds on the coefficients.

It would probably be more correct to call the Calderón formula a "method." It is based on a trick involving the Fourier transform.

We warn the reader that a small part of our discussion (not too much, we hope) will refer to the theory of distributions. The only concept we will use from it is that of "support." It will do no harm if the reader thinks of the support of a distribution as being the support of a measure—since, in our case, it will be.

Let $\psi \in \mathcal{C}_0^\infty(\mathbf{R}^d)$ be real, radial, have support contained in $\{x : |x| \leq 1\}$, satisfy $\int \psi \, dx = 0$, and be normalized so that, for all $\xi \in \mathbf{R}^d \setminus \{0\}$,

$$\int_0^\infty \left|\hat{\psi}(y\xi)\right|^2 \frac{dy}{y} = 1. \tag{5.1}$$

Formula 5.1, to put it mildly, calls for a few words of explanation. Since ψ is radial, so is $\hat{\psi}$. Thus, by a change of variable $u = y|\xi|$, we see that the left-hand side of 5.1 is independent of $\xi \neq 0$. Indeed, if we set $\xi \equiv (1, 0, 0, \ldots, 0)$, the left-hand side of 5.1 is simply

$$\int_0^\infty \left|\hat{\psi}(y, 0, 0, \ldots, 0)\right|^2 \frac{dy}{y}. \tag{5.2}$$

Since ψ is compactly supported and has integral 0, we have $|\hat{\psi}(y, 0, \ldots, 0)| \leq C_\psi y$ for small y; since ψ is smooth, we have $|\hat{\psi}(y, 0, \ldots, 0)| \leq C/y$ when y is large (strictly speaking, we have $\leq C/y^M$ for any M here, but $M = 1$

will do). Together these imply that the integral 5.2 is *finite* for any such ψ. Now we just take some non-trivial ψ and multiply it by an appropriate positive constant, to get a function satisfying 5.1.

Let's write $\mathbf{R}_+^{d+1} = \mathbf{R}^d \times (0, \infty)$. If $d = 1$, this is the familiar upper half-plane from complex analysis; we usually call it the upper half-*space*. Notice that $\mathbf{R}_+^{d+1}$ is an open subset of $\mathbf{R}^{d+1}$. This has the consequence that any compact subset of $\mathbf{R}_+^{d+1}$ has a strictly positive distance to $\partial \mathbf{R}_+^{d+1}$, which we think of as $\mathbf{R}^d$.

For $y > 0$, we let $\psi_y(t)$ stand for $y^{-d}\psi(t/y)$, the usual L^1-dilation. We claim that, if $f \in L^2(\mathbf{R}^d)$, then

$$\int |f|^2 \, dx = \int_{\mathbf{R}_+^{d+1}} |f * \psi_y(t)|^2 \, \frac{dt \, dy}{y}. \tag{5.3}$$

The proof is easy. By the Fourier Convolution Theorem, and by the Plancherel Theorem, the right-hand side of 5.3 is

$$\int_{\mathbf{R}^d} |\hat{f}(\xi)|^2 \left(\int_0^\infty |\hat{\psi}(y\xi)|^2 \, \frac{dy}{y} \right) d\xi$$

which equals $\int_{\mathbf{R}^d} |\hat{f}(\xi)|^2 \, d\xi$ because of 5.1. Equation 5.3 follows by another application of the Plancherel Theorem.

We now claim that, if $f \in L^2(\mathbf{R}^d)$, then

$$f = \int_{\mathbf{R}_+^{d+1}} (f * \psi_y(t)) \, \psi_y(x - t) \, \frac{dt \, dy}{y}$$

in some "useful" sense. Let us now make this sense precise! It is clear that, if $K \subset \mathbf{R}_+^{d+1}$ is any measurable set with compact closure contained in $\mathbf{R}_+^{d+1}$, then

$$f_{(K)}(x) \equiv \int_K (f * \psi_y(t)) \, \psi_y(x - t) \, \frac{dt \, dy}{y} \tag{5.4}$$

belongs to $L^2(\mathbf{R}^d)$ (because it's continuous and has compact support). We have a nice bound on $f_{(K)}$'s L^2 norm; to wit, $\|f_{(K)}\|_2 \leq \|f\|_2$, *independent of* K. To see this, let $h \in L^2(\mathbf{R}^d)$ satisfy $\|h\|_2 = 1$, and check the inner product $\int f_{(K)} \bar{h} \, dx$. It is

$$\int_K (f * \psi_y(t)) \, (\bar{h} * \psi_y(t))) \, \frac{dt \, dy}{y}.$$

By the Cauchy-Schwarz inequality, this integral is bounded in modulus by

$$\left(\int_{\mathbf{R}_+^{d+1}} |f * \psi_y(t)|^2 \, \frac{dt \, dy}{y} \right)^{1/2} \left(\int_{\mathbf{R}_+^{d+1}} |h * \psi_y(t)|^2 \, \frac{dt \, dy}{y} \right)^{1/2}.$$

(Notice that we have dropped K.) The first factor is no bigger than $\|f\|_2$ and the second equals 1. Thus, since h is arbitrary, $\|f_{(K)}\|_2 \leq \|f\|_2$.

We will say that $\{K_i\}_i$ is a *compact-measurable exhaustion* of $\mathbf{R}_+^{d+1}$ if each $K_1 \subset K_2 \subset K_3 \subset \cdots$ is an increasing sequence of measurable subsets of $\mathbf{R}_+^{d+1}$, such that each K_i has compact closure $\bar{K}_i$ contained in $\mathbf{R}_+^{d+1}$, and $\cup_i K_i = \mathbf{R}_+^{d+1}$. Note that we do *not* require any $\bar{K}_i$ to be contained in any K_{i+1}.

Remark. The only reason we require the K_i's to have compact closures is to ensure that the integrals defining $f_{(K_i)}$ will yield continuous, compactly supported functions.

The next theorem describes one sense in which the Calderón reproducing formula converges.

Theorem 5.1. *Let $f \in L^2(\mathbf{R}^d)$. If $\{K_i\}_i$ is any compact-measurable exhaustion of $\mathbf{R}_+^{d+1}$, then $f_{(K_i)} \to f$ weakly; i.e., $\int f_{(K_i)} \bar{h} \, dx \to \int f \bar{h} \, dx$ for all $h \in L^2(\mathbf{R}^d)$.*

But, with just a little more work, we can see that the convergence of the Calderón formula is actually much better.

Theorem 5.2. *Let $f \in L^2(\mathbf{R}^d)$. If $\{K_i\}_i$ is any compact-measurable exhaustion of $\mathbf{R}_+^{d+1}$, then $f_{(K_i)} \to f$ in the L^2 norm.*

Proof of Theorem 5.1. Because of equation 5.3, we have, for any $f \in L^2$,

$$\int_{\mathbf{R}^d} f(x) \, \bar{f}(x) \, dx = \int_{\mathbf{R}_+^{d+1}} (f * \psi_y(t)) \, (\bar{f} * \psi_y(t)) \, \frac{dt \, dy}{y}.$$

Notice that both integrals are absolutely convergent. By polarization (i.e., algebra), this implies that, for any f and h in L^2,

$$\int_{\mathbf{R}^d} f(x) \, \bar{h}(x) \, dx = \int_{\mathbf{R}_+^{d+1}} (f * \psi_y(t)) \, (\bar{h} * \psi_y(t)) \, \frac{dt \, dy}{y}; \qquad (5.5)$$

and, again, both integrals are absolutely convergent. Therefore, the left-hand side of 5.5 equals the limit of

$$\int_{K_i} (f * \psi_y(t)) \, (\bar{h} * \psi_y(t)) \, \frac{dt \, dy}{y} = \int_{\mathbf{R}^d} f_{(K_i)} \, \bar{h} \, dx.$$

Theorem 5.1 is proved.

Proof of Theorem 5.2. Let $f \in L^2$ and $\{K_i\}_i$ be as in the preceding theorem. For every m and n, $m < n$, $\|f_{(K_m)} - f_{(K_n)}\|_2$ is less than or equal to

$$\left(\int_{K_n \setminus K_m} |f * \psi_y(t)|^2 \, \frac{dt \, dy}{y} \right)^{1/2}.$$

(Repeat our earlier duality argument, but this time *don't* drop the K's.)
Because

$$\int_{\mathbf{R}^{d+1}_+} |f * \psi_y(t)|^2 \, \frac{dt\,dy}{y} < \infty,$$

this implies that $\|f_{(K_m)} - f_{(K_n)}\|_2 \to 0$ as $m \to \infty$, and so $\{f_{(K_i)}\}$ is Cauchy in L^2. Call its L^2 limit g. But the sequence $\{f_{(K_i)}\}$ converges weakly to f. Therefore $g = f$. That proves Theorem 5.2.

We have just shown that the linear mapping

$$f \mapsto Tf \equiv \int_{\mathbf{R}^{d+1}_+} (f * \psi_y(t)) \, \psi_y(x - t) \, \frac{dt\,dy}{y} \qquad (5.6)$$

(where the integral is defined as the L^2 limit of integrals over a compact-measurable exhaustion) is equal to the identity operator on L^2.

The statement that $Tf = f$, in the sense (and with the restrictions) we have described, is the Calderón reproducing formula.

Remark. Before saying more about this formula, we should notice that it comes with an unfortunate limitation: the equation $Tf = f$ only applies (for now) to functions $f \in L^2$. It turns out that this formula holds for all $f \in L^p$ $(1 < p < \infty)$, in precisely the same sense: If $K_1 \subset K_2 \subset K_3 \ldots \subset \mathbf{R}^{d+1}_+$ is any compact-measurable exhaustion of $\mathbf{R}^{d+1}_+$, then $f_{(K_i)} \to f$ in L^p. We will prove this after we develop some more theory. The reader should think of the Calderón formula as a continuous analogue of the result from exercise 2.11. The cornerstone of its proof will be a continuous analogue of the dyadic result $\|f\|_p \sim \|S(f)\|_p$.

The Calderón formula 5.6 looks a little like our earlier Haar function decomposition sum,

$$f = \sum_I \langle f, h_{(I)} \rangle h_{(I)};$$

and it should. The convolution $f * \psi_y(t)$ is an inner product between f and a suitably localized and normalized smooth function with cancelation, which is then multiplied by $\psi_y(x - t)$ (which is another suitably localized and normalized smooth function with cancelation), and summed (integrated) up. That's roughly what happened with the Haar functions.

We're about to make 5.6 look a lot more like a Haar function decomposition.

Definition 5.1. *If $Q \subset \mathbf{R}^d$ is a cube then*

$$T(Q) \equiv \{(x, y) \in \mathbf{R}^{d+1}_+ : \ x \in Q, \ \ell(Q)/2 \le y < \ell(Q)\}.$$

Analysts often refer to $T(Q)$ as the top half of the so-called *Carleson box* $\hat{Q} \equiv Q \times (0, \ell(Q))$. Before going further, let's observe some simple (and intentionally redundant) properties of $T(Q)$ and $\hat{Q}$.

1. If Q and Q' are cubes, then $Q \subset Q'$ if and only if $T(Q) \subset \hat{Q}'$, if and only if $\hat{Q} \subset \hat{Q}'$.

2. If Q and Q' are dyadic cubes and $Q \neq Q'$, then $T(Q) \cap T(Q') = \emptyset$.

3. If Q is a dyadic cube, then

$$\hat{Q} = \cup\{T(Q') : \ Q' \in \mathcal{D}_d, \ Q' \subset Q\},$$

and this is a disjoint union (by 2.).

4. The family $\{T(Q)\}_{Q \in \mathcal{D}_d}$ tiles $\mathbf{R}_+^{d+1}$.

Because of 4., we may cut up the integral in the Calderón reproducing integral as follows:

$$
\begin{aligned}
f &= \int_{\mathbf{R}_+^{d+1}} (f * \psi_y(t))\, \psi_y(x - t)\, \frac{dt\, dy}{y} \\
&= \sum_{Q \in \mathcal{D}_d} \int_{T(Q)} (f * \psi_y(t))\, \psi_y(x - t)\, \frac{dt\, dy}{y} \\
&\equiv \sum_{Q \in \mathcal{D}_d} b_{(Q)}(x).
\end{aligned}
$$

Our preceding arguments about compact-measurable exhaustions imply that this sum converges to f in L^2 in the following sense: If $\mathcal{F}_1 \subset \mathcal{F}_2 \subset \cdots$ are increasing, finite families of dyadic cubes such that $\cup_n \mathcal{F}_n = \mathcal{D}_d$, then

$$\sum_{Q \in \mathcal{F}_n} b_{(Q)} \to f$$

in L^2 as $n \to \infty$.

Now, if $t \in Q$ and $y \leq \ell(Q)$, then $\psi_y(x - t)$ can only be non-zero for x's in $\tilde{Q}$, the concentric triple of Q. It is also clear that

$$
\begin{aligned}
\int b_{(Q)}\, dx &= \int_{T(Q)} (f * \psi_y(t)) \left(\int \psi_y(x - t)\, dx \right) \frac{dt\, dy}{y} \\
&= \int_{T(Q)} (f * \psi_y(t))\, (0)\, \frac{dt\, dy}{y} = 0.
\end{aligned}
$$

We often express this fact by saying that $b_{(Q)}$ "inherits cancelation from ψ."

So, $b_{(Q)}$ is supported in $\tilde{Q}$ and has integral 0. How large is $b_{(Q)}$? The function $\psi_y(x - t)$ doesn't get any bigger than $c\, \ell(Q)^{-d}$ when $(t, y) \in T(Q)$. Therefore, by Hölder's Inequality and a little arithmetic (which the reader is encouraged to work through),

$$\|b_{(Q)}\|_\infty \leq c\lambda_Q(f)|Q|^{-1/2},$$

where we can take $\lambda_Q(f)$ to be

$$\left(\int_{T(Q)} |f * \psi_y(t)|^2 \, \frac{dt \, dy}{y} \right)^{1/2}. \tag{5.7}$$

Similarly, the gradient $\nabla_x \psi_y(x - t)$ has size no bigger than $c\ell(Q)^{-1-d}$ when $(t, y) \in T(Q)$, which, after we differentiate under the integral sign, yields

$$\|\nabla b_{(Q)}\|_\infty \le c\lambda_Q(f)\ell(Q)^{-1}|Q|^{-1/2}.$$

In other words, $b_{(Q)}$ is equal to $c\lambda_Q a_{(Q)}$, where $a_{(Q)}$ is adapted to $\tilde{Q}$, and $c > 0$ depends only on ψ (and, of course, d).

Putting it all together, we can decompose any $f \in L^2$ as

$$f = c \sum_{Q \in \mathcal{D}_d} \lambda_Q a_{(Q)},$$

where the sum converges in L^2, each $a_{(Q)}$ is adapted to $\tilde{Q}$, and the $\lambda_Q(f)$'s satisfy

$$\sum |\lambda_Q(f)|^2 = \int |f|^2 \, dx. \tag{5.8}$$

Let's define

$$\tilde{S}_{sd}(f)(x) \equiv \left(\sum_{Q \in \mathcal{D}_d} \frac{|\lambda_Q(f)|^2}{|\tilde{Q}|} \chi_{\tilde{Q}}(x) \right)^{1/2}. \tag{5.9}$$

(The reason for the $\tilde{Q}$'s will become clear in a moment.)

Equation 5.8 implies that $\|\tilde{S}_{sd}(f)\|_2 = \|f\|_2$ for $f \in L^2$. However, the coefficients $\lambda_Q(f)$ make sense for any $f \in L^1_{\text{loc}}(\mathbf{R}^d)$, and even for f's that are distributions (which the reader is free to think of as measures). We want to know: To what extent does $\tilde{S}_{sd}(f)$ control f's size in spaces other than $L^2(\mathbf{R}^d)$—in particular, weighted L^p spaces?

This is really several questions, which we will try to confront in a logical, intelligible order[1].

Using the Calderón formula, we can represent an arbitrary $f \in L^2$ as a limit of finite linear sums of adapted functions. We know how to use Littlewood-Paley theory to control such sums, *if they are indexed over good families of cubes*. Now, our sums are indexed over $\{\tilde{Q} : Q \in \mathcal{D}_d\}$, which is clearly not a good family.

Our first question is: How do we get around this?

Fortunately, the answer is not too bad.

If Q is any cube, and $m > 0$, we let mQ denote the cube having the same center and orientation as Q, and sidelength $m\ell(Q)$ (e.g., $\tilde{Q}$ is $3Q$). We set $m\mathcal{D}_d = \{mQ : Q \in \mathcal{D}_d\}$.

The answer to our first question is:

---

[1] "Logical" does not always imply "intelligible," nor vice versa.

Theorem 5.3. *Let m be an odd positive integer. The collection $m\mathcal{D}_d$ is equal to the union of m^d pairwise disjoint good families.*

Remark. When reading through this proof for the first time, the reader is advised to set m equal to 3. In fact, he's encouraged to do so, since we won't be needing it for higher values of m.

Proof of Theorem 5.3. It is sufficient to prove the theorem when $d = 1$. For the purposes of this proof, $[k]$ shall mean 'k mod m.' Since m is odd, 2 is invertible mod m; let p be its multiplicative inverse (mod m).

Every $I \in m\mathcal{D}$ has the form

$$I = [\frac{mj + s}{2^k}, \frac{m(j+1) + s}{2^k}), \tag{5.10}$$

where j, k, and s are integers, with $0 \le s < m$. For fixed k and s, denote the family of intervals satisfying 5.10 for some j by $\mathcal{G}_s^k$.

Now comes the fun part. If we divide $I \in \mathcal{G}_s^k$ in half, the resulting subintervals (check them!) belong to $\mathcal{G}_{[2s]}^{k+1}$. A slightly more detailed computation shows that every $I \in \mathcal{G}_s^k$ is either the right or left half of some $J \in \mathcal{G}_{[ps]}^{k-1}$ (the question of "right" or "left" depends on the parity of j). If we define, for $s = 0, 1, \ldots, m-1$,

$$\mathcal{G}_s \equiv \left(\cup_{k=0}^\infty \mathcal{G}_{[2^k s]}^k \right) \cup \left(\cup_{k=1}^\infty \mathcal{G}_{[p^k s]}^k \right),$$

then these are the required families.

Therefore, if $f = \sum_{Q \in \mathcal{D}_d} \lambda_Q a_{(Q)}$ is a finite linear sum of adapted functions, with each $a_{(Q)}$ adapted to $\tilde{Q}$, we can write $f = \sum_1^{3^d} f_j$, where each $f_j = \sum_{\tilde{Q} \in \mathcal{G}_j} \lambda_Q a_{(Q)}$, with the $\mathcal{G}_j$'s being good families. The methods of the previous chapter now let us control each f_j—and therefore f itself—with $\tilde{S}_{sd}(f)$. We just need to make sure that the hypotheses on our weights imply that appropriate inequalities hold for *all* the $\tilde{Q}$'s, $Q \in \mathcal{D}_d$. Usually we do this by asking that certain inequalities hold for all cubes Q, period.

One example should make our meaning clear.

Theorem 5.4. *Let $0 < p < \infty$ and let $p/2 < \eta$. Suppose that v and w are two weights such that, for all cubes $Q \subset \mathbf{R}^d$,*

$$\int_Q v(x) \left(\log(e + v(x)/v_Q) \right)^\eta dx \le \int_Q w(x) \, dx.$$

Then, if $f = \sum_{Q \in \mathcal{D}_d} \lambda_Q a_{(Q)}$ is a finite linear sum of adapted functions, with each $a_{(Q)}$ adapted to $\tilde{Q}$, we have

$$\int_{\mathbf{R}^d} |f|^p v \, dx \le C \int_{\mathbf{R}^d} (\tilde{S}_{sd}(f))^p w \, dx,$$

where $\tilde{S}_{sd}(f)$ is as defined in 5.9, and the constant C only depends on p, η, and d.

Of course, what goes for such finite sums of adapted functions $f = \sum_{Q \in \mathcal{D}_d} \lambda_Q a_{(Q)}$ goes just as well for functions in $L^2(\mathbf{R}^d)$; because, given $f \in L^2(\mathbf{R}^d)$, we can find a sequence of finite, nested subsets $\mathcal{F}_1 \subset \mathcal{F}_2 \subset \cdots$ such that $\cup \mathcal{F}_k = \mathcal{D}_d$ and the sequence of functions

$$f_{(k)} \equiv \sum_{Q \in \mathcal{F}_k} \lambda_Q a_{(Q)}$$

converges to f almost everywhere. The corresponding theorem for f will then follow by Fatou's Lemma.

So, arbitrary functions in $L^2(\mathbf{R}^d)$ are no problem.

What about the others?

That brings us to our second question: To what extent, and in what sense(s), does the Calderón formula represent f's in spaces other than $L^2(\mathbf{R}^d)$?

To answer it we need to step up and look more closely at the Calderón formula. When we do, we find that the formula is not so mysterious as it at first appeared.

Set $\eta = \psi * \psi$. We can rewrite 5.2 as

$$\int_0^\infty f * \eta_y(x) \, \frac{dy}{y},$$

which (formally) is the convolution of f with

$$\int_0^\infty \eta_y(x) \, \frac{dy}{y}, \tag{5.11}$$

whatever that means. Our preceding work in L^2 amounts to a case for saying that 5.11 gives a representation (in some sense) of δ_0, the Dirac mass at 0. Let's make this "in some sense" more precise.

Define

$$\phi(x) = \int_1^\infty \eta_y(x) \, \frac{dy}{y}.$$

It's not hard to see that the integral defining $\phi(x)$ converges absolutely, uniformly in x. A simple change of variable shows that, for any $t > 0$,

$$\phi_t(x) = \int_t^\infty \eta_y(x) \, \frac{dy}{y}.$$

So, the Calderón formula 5.2 can perhaps be reinterpreted as

$$f = \lim_{t \to 0+} f * \phi_t$$

or

$$f = \lim_{\substack{t \to 0+ \\ T \to \infty}} f * \phi_t - f * \phi_T,$$

both of which are true a.e. for $f \in L^p$ $(1 \le p < \infty)$, if ϕ is nice enough. (See exercise 5.7 at the end of this chapter.)

How "nice" is ϕ?

For $R > 1$ let's temporarily define

$$g_{(R)}(x) = \int_1^R \eta_y(x) \, \frac{dy}{y}.$$

It's clear that $g_{(R)} \in L^1$. Taking Fourier transforms, we get

$$\hat{g}_{(R)}(\xi) = \int_1^R |\hat{\psi}(y\xi)|^2 \, \frac{dy}{y}.$$

As $R \to \infty$, $\hat{g}_{(R)}(\xi) \nearrow$ to

$$F(\xi) \equiv \int_1^\infty |\hat{\psi}(y\xi)|^2 \, \frac{dy}{y}.$$

By Minkowski's Inequality for integrals:

$$\left(\int |F(\xi)|^2 \, d\xi \right)^{1/2} \le \int_1^\infty \left(\int |\hat{\psi}(y\xi)|^4 \, d\xi \right)^{1/2} \frac{dy}{y}$$

$$= \int_1^\infty \left(\int |\hat{\psi}(\xi)|^4 \, d\xi \right)^{1/2} \frac{dy}{y^{1+d/2}}$$

$$< \infty,$$

because ψ belongs to the Schwartz class. The Monotone Convergence Theorem now implies that the functions $\hat{g}_{(R)}$—and hence the functions $g_{(R)}$—have an L^2 limit. But we already know that the $g_{(R)}$'s converge *pointwise* to ϕ. Thus, $\phi \in L^2$, and its Fourier transform is what we've called $F(\xi)$:

$$\int_1^\infty |\hat{\psi}(y\xi)|^2 \, \frac{dy}{y}.$$

We can similarly show that $|\xi|^M F(\xi) \in L^2$ for all M, implying that ϕ is infinitely differentiable.

We claim that ϕ has *compact support*. What this means is that there is a compact set $K \subset \mathbf{R}^d$ such that $\phi(x) = 0$ almost everywhere on $\mathbf{R}^d \setminus K$. Indeed, we can take K to be $\{x \in \mathbf{R}^d : |x| \le 4\}$. We prove this by showing that, if $h \in \mathcal{C}_0^\infty(\mathbf{R}^d)$ has support disjoint from K, then $\int \phi h \, dx = 0$.

This last statement is easy to prove. Let $h \in \mathcal{C}_0^\infty(\mathbf{R}^d)$. If y is small, the integral $\int h(x)\eta_y(x) \, dx$ is bounded by a constant times y (exercise 5.2). Therefore the integral

$$\int_0^1 \int h(x)\eta_y(x) \, dx \, \frac{dy}{y} \tag{5.12}$$

converges. The mapping $\mathcal{C}_0^\infty(\mathbf{R}^d) \mapsto \mathbf{R}$ defined by 5.12 is a distribution with compact support; because, if the support of h is disjoint from $\{x \in \mathbf{R}^d : |x| \leq 4\}$, $h(x)\eta_y(x) \equiv 0$ for $0 < y \leq 1$. However, as a distribution, the function ϕ is equal to

$$\delta_0 - \int_0^1 \eta_y(x)\,\frac{dy}{y}.$$

In English, this last statement means that, for any $h \in \mathcal{C}_0^\infty(\mathbf{R}^d)$,

$$\int \phi(x)\,\bar{h}(x)\,dx = \bar{h}(0) - \int_0^1 \int \bar{h}(x)\eta_y(x)\,dx\,\frac{dy}{y}, \qquad (5.13)$$

which proves that ϕ has support contained inside K. The proof of 5.13 goes as follows:

$$\int \phi(x)\,\bar{h}(x)\,dx = \int \overline{\hat{h}(\xi)}\,\hat{\phi}(\xi)\,d\xi$$

$$= \int \overline{\hat{h}(\xi)}\left(1 - \int_0^1 |\hat{\psi}(y\xi)|^2\,\frac{dy}{y}\right)\,d\xi$$

$$= \int \overline{\hat{h}(\xi)}\,d\xi - \int \left(\int_0^1 \overline{\hat{h}(\xi)}\,|\hat{\psi}(y\xi)|^2\,\frac{dy}{y}\right)\,d\xi$$

$$= \bar{h}(0) - \int_0^1 \left(\int \overline{\hat{h}(\xi)}\,|\hat{\psi}(y\xi)|^2\,d\xi\right)\,\frac{dy}{y}$$

$$= \bar{h}(0) - \int_0^1 \int \bar{h}(x)\eta_y(x)\,dx\,\frac{dy}{y}.$$

The manipulations of the integrals are justified by the smoothness and rapid decay of h and ψ.

Now, knowing this, we can assert that ϕ is a smooth, radial function in L^1. The integral of ϕ is

$$F(0) = \lim_{\xi \to 0} \int_1^\infty |\hat{\psi}(y\xi)|^2\,\frac{dy}{y}$$

$$= \lim_{\xi \to 0} \int_{|\xi|}^\infty |\hat{\psi}(y(1,0,0,\ldots,0)))|^2\,\frac{dy}{y}$$

$$= \int_0^\infty |\hat{\psi}(y(1,0,0,\ldots,0))|^2\,\frac{dy}{y} = 1.$$

That makes ϕ nice enough for a lot of things. In the exercises we have outlined proofs of the following facts: 1) if $f \in L^p$ ($1 \leq p < \infty$), then it is the L^p and almost everywhere limit of $f * \phi_t$ as $t \to 0$; 2) if $1 < p < \infty$, then these are also true of the L^p limit of $f * \phi_t - f * \phi_T$ as $t \to 0$ and $T \to \infty$; 3) if $p = 1$, $f * \phi_t - f * \phi_T$ converges to f almost everywhere as $t \to 0$ and $T \to \infty$.

Assuming these, we now have almost all we need for a useful reinterpretation of the Calderón formula. What's missing is a justification of the equality:

$$f * \phi_t - f * \phi_T = \int_t^T f * \eta_y(x)\, \frac{dy}{y}.$$

Let's now provide this.

Note that both sides make sense for any $f \in L^p$. The problem in equating them comes from the fact that, on the left, we are essentially taking the difference of two improper integrals that might not converge in the sense of Lebesgue. Fortunately, this is easy to fix. The equation is easily seen to be true for $f \in \mathcal{C}_0^\infty(\mathbf{R}^d)$ and—here's the important part—both sides depend continuously on $f \in L^p$. Therefore the equation holds for *all* $f \in L^p$ ($1 \leq p < \infty$).

With this equation in hand, an L^p Littlewood-Paley estimate on f's size is straightforward.

For $k, n = 1, 2, \ldots$ and $f \in L^p$ ($1 \leq p < \infty$), set

$$f_{k,n} = \int_{2^{-n}}^{2^n} \int_{|t| \leq 2^k} \left(\int (f * \psi_y(t))\, \psi_y(x - t)\, dt \right) \frac{dy}{y}.$$

If $k_n \to \infty$ fast enough then, for all x, we will have $f_{k_n,n}(x) = f * \phi_{2^{-n}}(x) - f * \phi_{2^n}(x)$, for large enough n (depending on x, of course). For example: pick $k_n = 2^{2n}$. If $2^{-n} \leq y \leq 2^n$, then the support of ψ_y is contained in a ball of radius 2^n. Let $|x| \leq 2^r$, with $r > 0$. If $r < n$ then the ball centered at x and of radius 2^n will be contained in $\{t : |t| \leq 2^{2n}\}$, implying that

$$\begin{aligned}
f_{k_n,n}(x) &= \int_{2^{-n}}^{2^n} \int_{|t| \leq 2^{2n}} \left(\int (f * \psi_y(t))\, \psi_y(x - t)\, dt \right) \frac{dy}{y} \\
&= \int_{2^{-n}}^{2^n} \int_{\mathbf{R}^d} \left(\int (f * \psi_y(t))\, \psi_y(x - t)\, dt \right) \frac{dy}{y} \\
&= f * \phi_{2^{-n}}(x) - f * \phi_{2^n}(x).
\end{aligned}$$

Assume we've chosen such a sequence k_n. Then $f_{k_n,n} \to f$ almost everywhere. What's nice about this is that every $f_{k_n,n}$ can be written as a *finite* linear combination of adapted functions; not only that, but this sum is one for which we have good control on the sizes of the coefficients.

Let's call the region $\{(t, y) \in \mathbf{R}_+^{d+1} : |t| \leq 2^{k_n}, 2^{-n} \leq y \leq 2^n\}$ by the name R_n. Then

$$f_{k_n,n} = \int_{R_n} (f * \psi_y(t))\, \psi_y(x - t)\, \frac{dy\, dt}{y}.$$

For each $Q \in \mathcal{D}_d$ and each n, we can write

$$\lambda_{Q,n}(f_{k_n,n}) a_{(Q),n} \equiv \int_{R_n \cap T(Q)} (f * \psi_y(t))\, \psi_y(x - t)\, \frac{dy\, dt}{y},$$

where each $a_{(Q),n}$ is adapted to $\tilde{Q}$ and $|\lambda_{(Q),n}| \leq |\lambda_{(Q)}(f)|$. (The coefficients $\lambda_{(Q)}(f)$ are as defined by 5.7.) Note that, for each n, only finitely many $\lambda_{(Q),n}$'s are non-zero. Putting

$$f_{k_n,n} = \sum_{Q \in \mathcal{D}_d} \lambda_{Q,n}(f_{k_n,n}) a_{(Q),n}$$

gives the desired decomposition of $f_{k_n,n}$ as a linear sum of adapted functions.

The family $\{\tilde{Q} : Q \in \mathcal{D}_d\}$ is equal to the disjoint union of 3^d good families $\mathcal{G}_i$. Theorem 5.4 (with $v \equiv w \equiv 1$) can be applied to each of the functions

$$f_{k_n,n,i} \equiv \sum_{Q:\tilde{Q} \in \mathcal{G}_i} \lambda_{Q,n}(f_{k_n,n}) a_{(Q),n},$$

to obtain $(1 \leq p < \infty)$:

$$\int_{\mathbf{R}^d} |f_{k_n,n,i}|^p \, dx \leq C_{p,d} \int_{\mathbf{R}^d} (\tilde{S}_{sd,\mathcal{G}_i}(f_{k_n,n,i}))^p \, dx,$$

where

$$\tilde{S}_{sd,\mathcal{G}_i}(f_{k_n,n,i}) = \left(\sum_{Q:\tilde{Q} \in \mathcal{G}_i} \frac{|\lambda_{Q,n}(f_{k_n,n})|^2}{|\tilde{Q}|} \chi_{\tilde{Q}} \right)^{1/2}.$$

It is clear that each $\tilde{S}_{sd,\mathcal{G}_i}(f_{k_n,n,i}) \leq \tilde{S}_{sd}(f)$ pointwise (recall 5.9) and trivial that $f_{k_n,n} = \sum_i f_{k_n,n,i}$. Therefore, by Fatou's Lemma, if $f \in L^p(\mathbf{R}^d)$, $(1 \leq p < \infty)$, then

$$\int_{\mathbf{R}^d} |f|^p \, dx \leq C_{p,d} \int_{\mathbf{R}^d} (\tilde{S}_{sd}(f))^p \, dx. \tag{5.14}$$

This inequality generalizes readily to weighted spaces, and to L^p when $0 < p < 1$. If v and w are two weights satisfying the hypotheses of Theorem 5.4 and if f is the almost-everywhere limit of

$$\int_{R_n} (f * \psi_y(t)) \, \psi_y(x - t) \frac{dy \, dt}{y} \tag{5.15}$$

for *some* compact measurable exhaustion $R_n \nearrow \mathbf{R}_+^{d+1}$, then

$$\int_{\mathbf{R}^d} |f|^p \, v \, dx \leq C(p, d, \eta) \int_{\mathbf{R}^d} (\tilde{S}_{sd}(f))^p \, w \, dx,$$

where η is any number bigger than $p/2$.

As we have seen, if $f \in \cup_{1 \leq r < \infty} L^r$, then f can be written as the almost-everywhere limit of a sequence of functions of the form 5.15. We can sum up the meaning of much of this chapter in the following theorem.

Theorem 5.5. *Let $0 < p < \infty$, $\eta > p/2$, and suppose that v and w are two weights such that, for all cubes $Q \subset \mathbf{R}^d$,*

$$\int_Q v(x) \left(\log(e + v(x)/v_Q)\right)^\eta dx \leq \int_Q w(x) \, dx.$$

Suppose that $\psi \in \mathcal{C}_0^\infty(\mathbf{R}^d)$ is real, radial, has support contained in $\{x : |x| \leq 1\}$, has integral equal to 0, and also satisfies 5.1. There is a constant $C = C(\psi, p, d, \eta)$ such that, for all $f \in \cup_{1 \leq r < \infty} L^r$,

$$\int_{\mathbf{R}^d} |f|^p \, v \, dx \leq C_{p,d} \int_{\mathbf{R}^d} (\tilde{S}_{sd}(f))^p \, w \, dx, \tag{5.16}$$

where $\tilde{S}_{sd}(f)$ is as defined in 5.9. In particular, 5.16 holds if $v = w$ and $v \in A_\infty$.

Remark. Honesty compels us to point out that 5.14 and Theorem 5.5, while true, are slightly deceptive. It is now natural to jump to the conclusion that, if $f \in L^p(\mathbf{R}^d)$ $(1 < p < \infty)$, and $\{K_i\}$ is any compact-measurable exhaustion of $\mathbf{R}_+^{d+1}$, then $f_{(K_i)} \to f$ in L^p, with 5.14 holding. This is true, but we haven't proved it yet. We will do so in the next chapter.

The Calderón reproducing formula is a powerful tool, but we should not oversell it. It also has some powerful limitations. We have already seen one: The Calderón integral does NOT converge in the L^1 norm (although, properly interpreted, it does converge almost everywhere). In L^∞ the situation is even worse. If $f \in L^\infty(\mathbf{R}^d)$ we cannot, in general, expect the integral

$$\int_{\mathbf{R}_+^{d+1}} (f * \psi_y(t)) \, \psi_y(x - t) \, \frac{dt \, dy}{y}$$

to converge to f in *any* sense. For example, if f is a constant, the integrand is identically 0, which means that the Calderón formula cannot distinguish between two bounded functions that differ by a constant. (Fortunately, for bounded functions, this is the worst that can happen: see the exercises.)

As promised, we have shown that the Calderón formula allows us to write essentially arbitrary functions as limits of finite sums of adapted functions, and in ways that allow us to apply Littlewood-Paley analysis. We wish to make three final observations about these adapted function decompositions.

Observation 1: Our Littlewood-Paley estimates do not require very smooth adapted functions. Instead of asking that a function $a_{(Q)}$, adapted to a cube Q, satisfy

$$|a_{(Q)}(x) - a_{(Q)}(y)| \leq (|x - y|/\ell(Q))|Q|^{-1/2},$$

we could have asked that, for all x and y,

$$|a_{(Q)}(x) - a_{(Q)}(y| \leq (|x - y|/\ell(Q))^\alpha |Q|^{-1/2} \tag{5.17}$$

for some fixed $\alpha > 0$, independent of Q. What we are requiring is that $a_{(Q)}$
satisfy a suitably scaled Lipschitz (or Hölder) smoothness condition of order α.
The Littlewood-Paley estimates we obtained in chapter 4 for sums of adapted
functions also go through for these Hölder-smooth adapted functions, but with
constants that (of course) depend on α.

Nevertheless, it is sometimes useful to note:

Observation 2: The Calderón formula yields adapted functions with arbitrarily many derivatives. We obtained our adapted functions $a_{(Q)}$ from the
formula

$$\lambda_Q a_{(Q)} = \int_{T(Q)} (f * \psi_y(t))\, \psi_y(x - t)\, \frac{dt\, dy}{y}, \qquad (5.18)$$

where

$$\lambda_Q = c \left(\int_{T(Q)} |f * \psi_y(t)|^2 \, \frac{dt\, dy}{y} \right)^{1/2}.$$

However, since ψ is infinitely differentiable, we can take as many derivatives
as we like (in x) on both sides of 5.18. Simple estimates now imply that, for
all derivatives $D^\alpha(a_{(Q)})$, we will have

$$\|D^\alpha(a_{(Q)})\|_\infty \le C_{\alpha,d} \ell(Q)^{-|\alpha|} |Q|^{-1/2},$$

where $|\alpha|$ is the order of the differential operator D^α. We don't need this
extra smoothness for our basic Littlewood-Paley estimates, but it can come
in handy when studying the action of differential operators.

Observation 3: The Calderón formula yields adapted functions with arbitrarily many moments of cancelation. Let Δ be the Laplacian operator on $\mathbf{R}^d$,
and let ψ be as in 5.1. We could just as well write our Calderón representation
of f with respect to $C_M(\Delta^M \psi)$, where C_M is a constant chosen to ensure 5.1.
But $\Delta^M \psi$ is orthogonal to all polynomials of degree $< 2M$. Usually we don't
need this extra cancelation, but it can be useful when studying the action of
certain *integral* operators.

Exercises

5.1. The Calderón reproducing formula uses a function $\psi \in \mathcal{C}_0^\infty(\mathbf{R}^d)$ that is
real, radial, has support contained in $\{x : |x| \le 1\}$, satisfies $\int \psi\, dx = 0$, and
is not identically 0. Show that such a ψ exists.

5.2. Let h and η belong to $\mathcal{C}_0^\infty(\mathbf{R}^d)$, and suppose that $\int \eta\, dx = 0$. Show that
there is a constant C, depending only on h and η, so that, for all $0 < y \le 1$,
and all x, $|h * \eta_y(x)| \le Cy$.

5.3. Let ρ and ψ be two functions in $C_0^\infty(\mathbf{R}^d)$ having integrals equal to 0. Show that

$$\int_{\mathbf{R}_+^{d+1}} |\rho * \psi_y(t)| \frac{dt\,dy}{y} < \infty.$$

What does this say about the convergence of the Calderón reproducing formula for functions $\rho \in C_0^\infty(\mathbf{R}^d)$ that satisfy $\int \rho\,dx = 0$?

5.4. Let $f \in L^2(\mathbf{R}^d)$, and suppose that $E \subset \mathbf{R}_+^{d+1}$ is measurable. Let ψ be any function in $C_0^\infty(\mathbf{R}^d)$ satisfying $\int \psi\,dx = 0$. Show that, if $\{K_i\}$ is any compact-measurable exhaustion of $\mathbf{R}_+^{d+1}$, then

$$\lim_{i\to\infty} \int_{E\cap K_i} (f * \psi_y(t))\,\psi_y(x - t) \frac{dt\,dy}{y}$$

exists in the L^2 sense, and that the limit is independent of the sequence of sets $\{K_i\}$. Formally, this limit is equal to

$$f_{(E)}(x) \equiv \int_E (f * \psi_y(t))\,\psi_y(x - t) \frac{dt\,dy}{y}.$$

Show that $\|f_{(E)}\|_2 \le C_\psi \|f\|_2$. Show that this definition of $f_{(E)}$ coincides with our earlier formula 5.4 when E has compact closure contained in $\mathbf{R}_+^{d+1}$. Show that, if F and E are arbitrary measurable subsets of $\mathbf{R}_+^{d+1}$, and $E \subset F$, then $f_{(F)} - f_{(E)} = f_{(F\setminus E)}$.

5.5. This exercise builds on exercise 5.4. Let $f \in L^2(\mathbf{R}^d)$. Show that, for all $\epsilon > 0$, there is a compact $K \subset \mathbf{R}_+^{d+1}$ such that, for all measurable $E \subset \mathbf{R}_+^{d+1}$, if $K \subset E$, then $\|f_{(K)} - f_{(E)}\|_2 < \epsilon$.

5.6. Let $\psi \in C_0^\infty(\mathbf{R}^d)$ be real, radial, have support contained in $\{x : |x| \le 1\}$, satisfy $\int \psi\,dx = 0$, and be normalized so that, for all $\xi \ne 0$,

$$\int_0^\infty |\hat{\psi}(t\xi)|^2 \frac{dt}{t} = 1.$$

Let $\phi \in L^\infty(\mathbf{R}^d)$ satisfy $\phi * \psi_y(t) = 0$ for all $(t, y) \in \mathbf{R}_+^{d+1}$. Show that ϕ is a.e. constant. (Hint: First show that the conclusion follows if $\int \phi\rho\,dx = 0$ for all $\rho \in C_0^\infty(\mathbf{R}^d)$ with integral equal to 0. Then show that the hypothesis implies this.) Show that this conclusion can fail if ϕ is a polynomial.

5.7. Let $\phi \in C_0^\infty(\mathbf{R}^d)$ satisfy $\int \phi\,dx = 1$. This exercise guides you through a proof that, if $f \in L^p(\mathbf{R}^d)$ $(1 \le p < \infty)$, then

$$\lim_{\substack{T\to\infty \\ t\to 0}} (f * \phi_t(x) - \lim f * \phi_T(x)) = f(x)$$

almost everywhere, with convergence in L^p as well when $p > 1$.

a) Show that there is a constant C_ϕ such that, for all $f \in L^1_{\text{loc}}(\mathbf{R}^d)$,

$$\sup_{t>0} |f * \phi_t(x)| \le C_\phi M(f)(x).$$

(Hint: Assume that ϕ and f are non-negative, and look at exercise 2.7.)

b) Show that, if $f \in L^p(\mathbf{R}^d)$ and $1 \le p < \infty$, then

$$\lim_{T \to \infty} f * \phi_T(x) \to 0$$

uniformly. Show that this convergence holds in L^p if $p > 1$.

c) For $\delta > 0$ and $f \in L^1_{\text{loc}}(\mathbf{R}^d)$, define

$$\Omega_\delta(f)(x) \equiv \sup_{0 < t_1, t_2 < \delta} |f * \phi_{t_1}(x) - f * \phi_{t_2}(x)|,$$

and set $\Omega(f)(x) = \lim_{\delta \to 0} \Omega_\delta(f)(x)$. Notice these facts about $\Omega(f)$: it might be infinite, but it is defined everywhere and it is measurable; $\Omega(f + g) \le \Omega(f) + \Omega(g)$; $\Omega(f) \le 2C_\phi M(f)$; $\Omega(f)(x) = 0$ if and only if $\lim_{t \to 0} f * \phi_t(x)$ exists; if f is continuous, then $\Omega(f)(x) = 0$ everywhere.

d) Show that, if f is continuous and with compact support, then

$$\lim_{t \to 0} f * \phi_t(x) \to f(x)$$

uniformly and in L^p ($0 < p \le \infty$). Use this to infer that $f * \phi_t \to f$ in L^p for arbitrary $f \in L^p$ ($1 \le p < \infty$).

e) Let $f \in L^p$ ($1 \le p < \infty$), and write $f = g + b$, where g is continuous with compact support, and $\|b\|_p < \epsilon$. Put some of the preceding estimates together to infer that, for all $\eta > 0$, the set $\{x : \Omega(f)(x) > \eta\}$ has measure zero, and therefore that $\lim_{t \to 0} f * \phi_t(x)$ exists almost everywhere.

f) Last step: under the same assumptions as in step e), show that the almost-everywhere value of $\lim_{t \to 0} f * \phi_t(x)$ is $f(x)$.

Notes

The original form of the Calderón reproducing formula appears in [4]. Versions of it more closely resembling what we see in this chapter appear in [6], [7], [5], [9], and [57]. Our proofs of Theorem 5.1 and Theorem 5.2 are heavily indebted to [53]; see also [15] for one treatment of the L^2 case. The papers [57] and [9] exploit the connection between Calderón's formula and adapted function decompositions essentially as we do here. The combinatorial argument (Theorem 5.3) used to construct "good" families, which is presented here, is from [62], but is similar to constructions in [10] and [25]. The convergence of the Calderón formula for smooth functions with good decay is proved in [22]. The use of the Calderón formula, combined with dyadic results, to prove weighted Littlewood-Paley inequalities for "arbitrary" functions, is based on the treatment in [62].

The Calderón Reproducing Formula II

The way we have been discussing the square function, in terms of Haar functions and adapted functions, is not how it has traditionally been done. Now we want to describe how Littlewood-Paley theory, as we have developed it, relates to a more classical theory.

We wish to emphasize that we will not be treating the classical theory of the square function in depth. Instead, we will develop some "semi-classical" (real-variable) theory and show how it meshes with some familiar[1] square function results. Our reasons for this are two-fold: 1) it saves space; and, 2) the real-variable theory is closer to the spirit of modern work in analysis.

Let $\psi \in \mathcal{C}_0^\infty(\mathbf{R}^d)$ be real, radial, have support contained in $\{x : |x| \le 1\}$, satisfy $\int \psi \, dx = 0$, and be normalized so that, for all $\xi \in \mathbf{R}^d \setminus \{0\}$,

$$\int_0^\infty |\hat{\psi}(y\xi)|^2 \, \frac{dy}{y} = 1.$$

This is equation 5.1 from the preceding chapter. For $\alpha > 0$ and $x \in \mathbf{R}^d$, let $\Gamma_\alpha(x)$ be the "cone" over x:

$$\Gamma_\alpha(x) \equiv \{(t, y) \in \mathbf{R}_+^{d+1} : |x - t| < \alpha y\}.$$

This cone is said to have aperture equal to α. We call $\Gamma_1(x)$ the "standard cone," and we usually denote it by plain $\Gamma(x)$. Notice that, if $0 < \alpha < \beta$, then $\Gamma_\alpha(x) \subset \Gamma_\beta(x)$.

With α and ψ fixed, we define the (real-variable) square function $S_{\psi,\alpha}(f)$ of a function $f \in L_{\mathrm{loc}}^1(\mathbf{R}^d)$ by

$$S_{\psi,\alpha}(f)(x) \equiv \left(\int_{\Gamma_\alpha(x)} |f * \psi_y(t)|^2 \, \frac{dt \, dy}{y^{d+1}} \right)^{1/2}. \tag{6.1}$$

---

[1] Familiar to older mathematicians!

In analogy with our definition of $\Gamma_\alpha(x)$, we will denote $S_{\psi,1}(f)$ by plain $S_\psi(f)$.

An analogous operator, the (real-variable) g-function, is defined by the equation:

$$g_\psi(f)(x) \equiv \left(\int_0^\infty |f * \psi_y(x)|^2 \frac{dy}{y} \right)^{1/2} . \tag{6.2}$$

These operators are sublinear and non-negative. They are also bounded on L^2. Indeed, for any $\alpha > 0$,

$$\int_{\mathbf{R}^d} (S_{\psi,\alpha}(f)(x))^2 \, dx = \int_{\mathbf{R}_+^{d+1}} |f * \psi_y(t)|^2 \, |x : \ (t,y) \in \Gamma_\alpha(x)| \, \frac{dt \, dy}{y^{d+1}}$$

$$= C_d \int_{\mathbf{R}_+^{d+1}} |f * \psi_y(t)|^2 \, \alpha^d y^d \, \frac{dt \, dy}{y^{d+1}}$$

$$= C_d \alpha^d \int_{\mathbf{R}_+^{d+1}} |f * \psi_y(t)|^2 \, \frac{dt \, dy}{y}$$

$$= C_d \alpha^d \|f\|_2,$$

when $f \in L^2$; and the analogous computation for $g_\psi(f)$ is even simpler. Notice that, up to possible constant factors, $S_{\psi,\alpha}$ and $g_\psi(f)$ induce nonlinear isometries on L^2.

I claim that we have already seen the operator $S_{\psi,\alpha}(f)$, though in a disguised form. Recall our definition of $\tilde{S}_{sd}(f)$,

$$\tilde{S}_{sd}(f)(x) \equiv \left(\sum_{Q \in \mathcal{D}_d} \frac{|\lambda_Q(f)|^2}{|\tilde{Q}|} \chi_{\tilde{Q}}(x) \right)^{1/2} ,$$

where

$$|\lambda_Q(f)| = \left(\int_{T(Q)} |f * \psi_y(t)|^2 \frac{dt \, dy}{y} \right)^{1/2} .$$

The function $\tilde{S}_{sd}(f)$ is pointwise comparable to

$$\left(\sum_{\substack{Q \in \mathcal{D}_d \\ x \in \tilde{Q}}} \int_{T(Q)} |f * \psi_y(t)|^2 \frac{dt \, dy}{y^{d+1}} \right)^{1/2} .$$

There is a positive constant $\alpha_1(d)$ such that, if $(t,y) \in T(Q)$ and $x \in \tilde{Q}$, then $|x - t| < \alpha_1(d)y$; conversely, there is another positive constant, $\alpha_2(d)$, such that, if $|x - t| < \alpha_2(d)y$, then (t,y) lies in a $T(Q)$ such that $x \in \tilde{Q}$. Put together, these say that $\tilde{S}_{sd}(f)$ is less than or equal to a constant times $S_{\psi,\alpha_1(d)}(f)$, and that it is larger than or equal to a constant times $S_{\psi,\alpha_2(d)}(f)$. Therefore, if

$$\int_{\mathbf{R}^d} |f|^p \, v \, dx \le C \int_{\mathbf{R}^d} (\tilde{S}_{sd}(f))^p \, w \, dx$$

holds for all f in some test class, it automatically holds (with a different constant C) if we replace $\tilde{S}_{sd}(f)$ with $S_{\psi,\alpha_1(d)}(f)$; and, likewise, if

$$\int_{\mathbf{R}^d} (\tilde{S}_{sd}(f))^p \, v \, dx \le C \int_{\mathbf{R}^d} |f|^p \, w \, dx$$

holds for all suitable f, it will hold with $S_{\psi,\alpha_2(d)}(f)$ in place of $\tilde{S}_{sd}(f)$; and, of course, weighted norm inequalities involving $S_{\psi,\alpha_1(d)}(f)$ and $S_{\psi,\alpha_2(d)}(f)$ automatically imply others for $\tilde{S}_{sd}(f)$.

The problem then arises of how to relate the L^p norms of $S_{\psi,\alpha_1(d)}(f)$ and $S_{\psi,\alpha_2(d)}(f)$ if we are working in weighted spaces or if $p \ne 2$.

We will avoid many technical problems in studying these objects if we define a new square function, one which is independent of any particular kernel ψ. This square function pointwise dominates all of the $S_{\psi,\alpha}(f)$'s and $g_\psi(f)$'s, but is not essentially larger, or any harder to handle, than any particular $S_{\psi,\alpha}(f)$. Using it, we will be able to show with relatively little trouble that $S_{\psi,\alpha}(f)$ and $g_\psi(f)$ have $L^p(\mathbf{R}^d)$ norms comparable to $\|f\|_p$ when $1 < p < \infty$. This new square function also has the virtue of being essentially aperture independent: its values for one aperture are pointwise comparable to its values for any other aperture, and this pointwise comparability even holds good for its "zero-aperture" version. This property makes it well-suited for work in weighted spaces.

For $\beta > 0$, let $\mathcal{C}_\beta$ be the family of functions $\phi : \mathbf{R}^d \mapsto \mathbf{R}$ having their supports contained in $\{x : |x| \le 1\}$, satisfying $\int \phi \, dx = 0$, and such that, for all x and x', $|\phi(x) - \phi(x')| \le |x - x'|^\beta$. If $f \in L^1_{\mathrm{loc}}(\mathbf{R}^d)$ and $(t, y) \in \mathbf{R}_+^{d+1}$, we define

$$A_\beta(f)(t, y) \equiv \sup_{\phi \in \mathcal{C}_\beta} |f * \phi_y(t)|.$$

Notice that $A_\beta(f)(t, y)$ is a measurable function of (t, y) and, for every fixed $x \in \mathbf{R}^d$, $A_\beta(f)(x, y)$ is a measurable function of y. These imply that the following definitions make sense.

Definition 6.1. *Given the family $\mathcal{C}_\beta$, the intrinsic square function of f, of aperture α, is defined by the equation*

$$G_{\alpha,\beta}(f)(x) \equiv \left(\int_{\Gamma_\alpha(x)} (A_\beta(f)(t, y))^2 \, \frac{dt \, dy}{y^{d+1}} \right)^{1/2}.$$

When $\alpha = 1$, we write $G_{\alpha,\beta}(f)$ simply as $G_\beta(f)$. The intrinsic g-function of f is defined by the equation

$$g_\beta(f)(x) \equiv \left(\int_0^\infty (A_\beta(f)(x, y))^2 \, \frac{dy}{y} \right)^{1/2}.$$

The discretized intrinsic g-function *of f is defined by the equation*

$$\sigma_\beta(f)(x) \equiv \left(\sum_{-\infty}^{\infty} (A_\beta(f)(x, 2^k))^2 \right)^{1/2}.$$

It is trivial that $S_{\psi,\alpha}(f) \le C(\psi, \beta, d)G_{\alpha,\beta}(f)$ pointwise, and therefore that $\|f\|_2 \le C\|G_{\alpha,\beta}(f)\|_2$ when $f \in L^2$. Any inequality of the form

$$\int_{\mathbf{R}^d} |f(x)|^p \, v \, dx \le C \int_{\mathbf{R}^d} (S_{\psi,\alpha}(f)(x))^p \, w \, dx$$

for any $0 < p < \infty$ and any weights v and w, automatically implies

$$\int_{\mathbf{R}^d} |f(x)|^p \, v \, dx \le C \int_{\mathbf{R}^d} (G_{\alpha,\beta}(f)(x))^p \, w \, dx. \qquad (6.3)$$

For example, our work in the preceding chapter shows that 6.3 holds for any $0 < p < \infty$, any $f \in \cup_{1 \le r < \infty} L^r$, and any pair of weights v and w such that

$$\int_Q v(x) \, (\log(e + v(x)/v_Q))^\eta \, dx \le \int_Q w(x) \, dx$$

holds for all cubes Q, where $\eta > p/2$.

Only slightly less trivial is the fact that if α and α' are two aperture sizes, then $G_{\alpha,\beta}(f)$ and $G_{\alpha',\beta}(f)$ are pointwise comparable. But even more is true. It is not hard to show that $g_\beta(f)$, $\sigma_\beta(f)$, and $G_\beta(f)$ are all pointwise comparable. We refer the reader to the exercises for the ideas behind the proofs of these facts; we will take them for granted.

Our next theorem is:

Theorem 6.1. *Let $1 < p \le 2$ and $\beta > 0$. There is a constant $C = C(p, d, \beta)$ such that, for all $f \in L^1_{\text{loc}}(\mathbf{R}^d)$ and all weights v,*

$$\int_{\mathbf{R}^d} (G_\beta(f))^p \, v \, dx \le C \int_{\mathbf{R}^d} |f|^p \, M(v) \, dx.$$

The meaning of Theorem 6.1 is that, while the intrinsic square function dominates all the square functions $S_{\psi,\alpha}(f)$ and $g_\psi(f)$, it is not much bigger than any of them. Of course, Theorem 6.1 immediately implies $\|S_{\psi,\alpha}(f)\|_p + \|g_\psi(f)\|_p \le C(p, \alpha, \psi, d)\|f\|_p$ for all $1 < p < \infty$, all ψ, and all α.

Proof of Theorem 6.1. The proof has three main steps.

Step 1.

$$\|G_\beta(f))\|_2 \le C(d, \beta)\|f\|_2.$$

Step 1 is the key to the proof.

Step 2.

$$\int_{\mathbf{R}^d} (G_\beta(f))^2 \, v \, dx \le C(d, \beta) \int_{\mathbf{R}^d} |f|^2 \, M(v) \, dx$$

for all f and all weights v.

Step 3. There is a constant $C = C(d, \beta)$ so that, for all f, all weights v, and all $\lambda > 0$,

$$v\left(\{x : \ G_\beta(f)(x) > \lambda\}\right) \le \frac{C}{\lambda} \int |f| \, M(v) \, dx.$$

Theorem 6.1 will then follow by interpolation.

Step 1. The L^2 boundedness of $G_\beta(f)$ is equivalent to having

$$\int_{\mathbf{R}_+^{d+1}} (A_\beta(f)(t, y))^2 \frac{dt \, dy}{y} \le C(\beta, d) \int_{\mathbf{R}^d} |f|^2 \, dx$$

for all $f \in L^2$, which is the same as showing

$$\int_K (A_\beta(f)(t, y))^2 \frac{dt \, dy}{y} \le C(\beta, d) \int_{\mathbf{R}^d} |f|^2 \, dx \qquad (6.4)$$

for all compact $K \subset \mathbf{R}_+^{d+1}$. Inequality 6.4 is what we will show.

Let $K \subset \mathbf{R}_+^{d+1}$ be compact and suppose that $\|f\|_2 \le 1$. If $\psi \in \mathcal{C}_\beta$ then $f * \psi_y(t)$ is continuous on K. The function $A_\beta(f)(t, y)$ is also continuous on K. Therefore we can choose, in a measurable (indeed, piecewise constant) fashion, functions $\psi^{(t,y)} \in \mathcal{C}_\beta$ such that, for every $(t, y) \in K$,

$$|f * \psi_y^{(t,y)}(t)| \ge (1/2) A_\beta(f)(t, y).$$

Let $g : \mathbf{R}_+^{d+1} \mapsto \mathbf{R}$ be a measurable function such that

$$\int_K |g(t, y)|^2 \frac{dt \, dy}{y} = 1.$$

We will be done if we can show

$$\left| \int_K (f * \psi_y^{(t,y)}(t)) \, g(t, y) \frac{dt \, dy}{y} \right| \le C(\beta, d). \qquad (6.5)$$

The integral in 6.5 is equal to

$$\int_{\mathbf{R}^d} f(x) \left(\int_K g(t, y) \, \psi_y^{(t,y)}(t - x) \frac{dt \, dy}{y} \right) dx.$$

Set

$$G(x) \equiv \int_K g(t, y) \, \psi_y^{(t,y)}(t - x) \frac{dt \, dy}{y}.$$

Since the integration is over a compact set, this is a bounded, continuous function of x, with bounded support. What we need to show is that $\|G\|_2 \leq C(\beta, d)$. We will now do this.

We can rewrite the integral defining G as

$$\sum_Q \int_{K \cap T(Q)} g(t, y)\, \psi_y^{(t,y)}(t - x)\, \frac{dt\, dy}{y};$$

and, as our earlier work has shown, this is in fact a finite sum. We can rewrite each summand as

$$\int_{K \cap T(Q)} g(t, y)\, \psi_y^{(t,y)}(t - x)\, \frac{dt\, dy}{y} = \lambda_Q a_{(Q)},$$

where

$$\lambda_Q = \left(\int_{K \cap T(Q)} |g(t, y)|^2\, \frac{dt\, dy}{y} \right)^{1/2}$$

and $a_{(Q)}$ is a continuous function having certain nice properties. In particular:

1. The support of $a_{(Q)}$ is contained inside $\tilde{Q}$, the triple of Q.
2. $\int a_{(Q)}\, dx = 0$.
3. There is a constant $C(\beta, d)$ such that, for all x and x',

$$|a_{(Q)}(x) - a_{(Q)}(x')| \leq C(\beta, d)(|x - x'|/\ell(Q))^\beta |Q|^{-1/2}.$$

Each of these properties follows from a corresponding defining property for $\mathcal{C}_\beta$. The only moderately tricky one is number 3, for which the crucial inequality to show is:

$$\left(\int_{T(Q)} \left| \psi_y^{(t,y)}(t - x) - \psi_y^{(t,y)}(t - x') \right|^2 \frac{dt\, dy}{y} \right)^{1/2} \leq$$
$$C(\beta, d)(|x - x'|/\ell(Q))^\beta |Q|^{-1/2},$$

which is easy, but we should note how important it is that the smoothness property of functions in $\mathcal{C}_\beta$ holds pointwise, and not merely in an averaged sense.

In other words, the functions $a_{(Q)}$ are, up to a positive multiple, *adapted functions*, adapted to the dyadic triples $\tilde{Q}$. Our earlier work[2] implies that, for every finite linear sum $\sum_Q \gamma_Q a_{(Q)}$,

$$\int |\sum_Q \gamma_Q a_{(Q)}|^2\, dx \leq C(\beta, d) \int \left(\sum_Q \frac{|\gamma_Q|^2}{|Q|} \chi_{\tilde{Q}} \right) dx = C(\beta, d) \sum_Q |\gamma_Q|^2.$$
$$(6.6)$$

---

[2] Apply Theorem 5.3 to the family of $\tilde{Q}$'s, and then think: Theorem 4.6 with $v \equiv 1$.

Applying this inequality to the finite linear sum $G = \sum_Q \lambda_Q a_{(Q)}$ yields

$$\int |G(x)|^2\, dx \leq C(\beta, d) \sum_Q |\lambda_Q|^2$$

$$= C(\beta, d) \int_K |g(t, y)|^2\, \frac{dt\, dy}{y},$$

which is what we wanted for Step 1[3].

Step 2. Let v be an arbitrary non-negative function in $L^1_{\text{loc}}(\mathbf{R}^d)$. For $(t, y) \in \mathbf{R}^{d+1}_+$, let $B(t; y)$ be the ball $\{x \in \mathbf{R}^d : |x - t| < y\}$. By Fubini-Tonelli,

$$\int_{\mathbf{R}^d} (G_\beta(f))^2\, v\, dx = C(d) \int_{\mathbf{R}^{d+1}_+} (A_\beta(f)(t, y))^2\, \frac{v(B(t; y))}{|B(t; y)|}\, \frac{dt\, dy}{y}.$$

For each integer k, define

$$F^k = \{(t, y) \in \mathbf{R}^{d+1}_+ : 2^k < \frac{v(B(t; y))}{|B(t; y)|} \leq 2^{k+1}\}.$$

These sets are disjoint and their union is $\{(t, y) \in \mathbf{R}^{d+1}_+ : v(B(t; y)) > 0\}$. Therefore

$$\int_{\mathbf{R}^d} (G_\beta(f))^2\, v\, dx = C(d) \sum_k \int_{F^k} (A_\beta(f)(t, y))^2\, \frac{v(B(t; y))}{|B(t; y)|}\, \frac{dt\, dy}{y}$$

$$\leq C(d) \sum_k 2^k \int_{F^k} (A_\beta(f)(t, y))^2\, \frac{dt\, dy}{y}.$$

Let's observe two things about the sets F^k: a) if $(t, y) \in F^k$ then $B(t; y) \subset \{x \in \mathbf{R}^d : M(v) > 2^k\} \equiv E_k$; b) if $(t, y) \in F^k$ then $A_\beta(f)(t, y) = A_\beta(f\chi_{E_k})(t, y)$. The first fact is trivial. The second follows because, if $\psi \in \mathcal{C}_\beta$, the function $\psi_y(t - \cdot)$ has support contained in $B(t; y)$. Therefore, for each k,

$$\int_{F^k} (A_\beta(f)(t, y))^2\, \frac{dt\, dy}{y} = \int_{F^k} (A_\beta(f\chi_{E_k})(t, y))^2\, \frac{dt\, dy}{y}$$

$$\leq C(\beta, d) \int |f\chi_{E_k}|^2\, dx$$

$$= C(\beta, d) \int |f|^2\, \chi_{E_k}\, dx,$$

---

[3] Referring to "our earlier work" to justify 6.6 is really a bit of overkill. For a more direct proof, see exercise 6.13 at the end of this chapter.

where the second line follows from Step 1. Summing on k, we now get

$$\int_{\mathbf{R}^d} (G_\beta(f))^2 \, v \, dx \le C(\beta, d) \sum_k 2^k \int |f|^2 \chi_{E_k} \, dx$$

$$= C(\beta, d) \int_{\mathbf{R}^d} |f|^2 \left(\sum_k 2^k \chi_{E_k} \right) dx$$

$$\le C(\beta, d) \int_{\mathbf{R}^d} |f|^2 \, M(v) \, dx,$$

which finishes Step 2.

Step 3. We will only prove Step 3 for functions f that have the following property: for every $\epsilon > 0$, there is an N such that, if $Q \subset \mathbf{R}^d$ is any cube with $\ell(Q) > N$, then

$$\frac{1}{|Q|} \int_Q |f| \, dx < \epsilon.$$

(One way to remove this restriction is outlined in an exercise.) That said, let $\{Q_k^\lambda\}_k$ be the maximal dyadic cubes such that

$$\frac{1}{|Q_k^\lambda|} \int_{Q_k^\lambda} |f| \, dx > \lambda.$$

Each Q_k^λ is contained in the set where $M_d(f) > \lambda$, but in fact a little more is true. If we let $\tilde{Q}_k^\lambda$ denote the triple of Q_k^λ, then $\tilde{Q}_k^\lambda$ is contained in the set where $M(f) > c\lambda$, where c is a positive constant depending only on d (note that this second maximal function does not have a subscript 'd'). This is because, if $x \in \tilde{Q}_k^\lambda$, there is a ball (or cube: it doesn't matter which) not much bigger than Q_k^λ, which contains both x and Q_k^λ.

We write $f = g + b$, where, following a familiar pattern,

$$g = \begin{cases} f_{Q_k^\lambda} & \text{if } x \in Q_k^\lambda; \\ f(x) & x \notin \cup_k Q_k^\lambda; \end{cases}$$

and $b = \sum_k b_k$, with $b_k \equiv (f - f_{Q_k^\lambda}) \chi_{Q_k^\lambda}$.

Set $\Omega = \{x : M(f)(x) > c\lambda\}$, where c is the positive constant from a few lines ago. According to exercise 3.5, there is a constant $C = C(d)$ such that

$$v(\Omega) \le \frac{C}{\lambda} \int |f| \, M(v) \, dx.$$

Therefore it will be enough to show that

$$v(\{x \notin \Omega : G_\beta(f)(x) > \lambda\}) \le \frac{C}{\lambda} \int |f| \, M(v) \, dx,$$

which will follow from

$$v\left(\{x \notin \Omega : \; G_\beta(g)(x) > \lambda/2\}\right) \leq \frac{C}{\lambda} \int |f| \, M(v) \, dx \tag{6.7}$$

and

$$v\left(\{x \notin \Omega : \; G_\beta(b)(x) > \lambda/2\}\right) \leq \frac{C}{\lambda} \int |f| \, M(v) \, dx. \tag{6.8}$$

The proof of 6.7 will depend on our Step 2 result and on a lemma, whose proof we leave to the reader:

Lemma 6.1. *Let Q be a cube, and set $S = \mathbf{R}^d \setminus \tilde{Q}$ (note the tilde). There is a constant C, depending only on d, so that, for any weight v,*

$$\sup_{x \in Q} M(v\chi_S)(x) \leq C \inf_{x \in Q} M(v)(x).$$

By the Step 2 inequality,

$$\int_{\mathbf{R}^d \setminus \Omega} (G_\beta(g))^2 \, v \, dx \leq C \int |g|^2 \, M(v\chi_{\mathbf{R}^d \setminus \Omega}) \, dx$$

$$\leq C \int_{\mathbf{R}^d \setminus \cup Q_k^\lambda} |f|^2 \, M(v) \, dx + C \sum_k |f_{Q_k^\lambda}|^2 \int_{Q_k^\lambda} M(v\chi_{\mathbf{R}^d \setminus \Omega}) \, dx$$

$$\leq C\lambda \int_{\mathbf{R}^d} |f| \, M(v) \, dx + C\lambda \sum_k |f_{Q_k^\lambda}| \int_{Q_k^\lambda} M(v\chi_{\mathbf{R}^d \setminus \Omega}) \, dx,$$

where the last line (factoring out the λ's) follows from the usual bound on g. Lemma 6.1 implies that, for each Q_k^λ,

$$|f_{Q_k^\lambda}| \int_{Q_k^\lambda} M(v\chi_{\mathbf{R}^d \setminus \Omega}) \, dx \leq C \int_{Q_k^\lambda} |f| \, M(v) \, dx.$$

Therefore,

$$\int_\Omega (G_\beta(g))^2 \, v \, dx \leq C\lambda \int_{\mathbf{R}^d} |f| \, M(v) \, dx + C\lambda \sum_k \int_{Q_k^\lambda} |f| \, M(v) \, dx$$

$$\leq C\lambda \int |f| \, M(v) \, dx.$$

The weak-type estimate for g now follows from Chebyshev's Inequality (i.e., dividing both sides by $(\lambda/2)^2$).

To estimate 6.8, we will show that, for each k,

$$\int_{\mathbf{R}^d \setminus \Omega} G_\beta(b_k) \, v \, dx \leq C(\alpha, d) \int_{Q_k^\lambda} |f| \, M(v) \, dx. \tag{6.9}$$

Summing on k will then yield

$$\int_{\mathbf{R}^d \setminus \Omega} G_\beta(b) \, v \, dx \leq C(\alpha, d) \sum_k \int_{Q_k^\lambda} |f| \, M(v) \, dx,$$

from which the weak $(1, 1)$ estimate will follow at once.

Inequality 6.9 is a direct consequence of a simple fact: Suppose that $h \in L^1(Q)$, Q a cube, and $\int_Q h\,dx = 0$. If $d(x, Q)$ means the distance from x to Q, then, for all x such that $d(x, Q) > \ell(Q)$,

$$G_\beta(h)(x) \le C(\beta, d)\|h\|_1 |Q|^{-1}(1 + |x - x_Q|/\ell(Q))^{-d-\beta}, \qquad (6.10)$$

where we are using x_Q to denote Q's center. This implies 6.9, because it yields

$$\int_{\mathbf{R}^d \setminus \Omega} G_\beta(b_k)\, v\,dx \le$$
$$C(\beta, d)\left(\int_{Q_k^\lambda} |f|\,dx\right)\left(|Q_k^\lambda|^{-1}\int_{\mathbf{R}^d} v(x)\,(1 + |x - x_{Q_k^\lambda}|/\ell(Q_k^\lambda))^{-d-\beta}\,dx\right).$$

But

$$|Q_k^\lambda|^{-1}\int_{\mathbf{R}^d} v(x)\,(1 + |x - x_{Q_k^\lambda}|/\ell(Q_k^\lambda))^{-d-\beta}\,dx \le C(\alpha, d)\inf_{x \in Q_k^\lambda} M(v)(x)$$

(see exercise 2.7), and therefore

$$\int_{\mathbf{R}^d \setminus \Omega} G_\beta(b_k)\, v\,dx \le C(\beta, d)\int_{Q_k^\lambda} |f|\,M(v)\,dx.$$

We finish by proving 6.10.

Suppose $d(x, Q) > \ell(Q)$, $(t, y) \in \Gamma(x)$, and $\phi \in \mathcal{C}_\beta$. Since ϕ is supported inside $\{x : |x| \le 1\}$ and $|x - t| < y$, the convolution $h * \phi_y(t)$ will be zero unless $y > c'|x - x_Q|$, where c' is some positive constant that depends on d. We merely note this fact now; it will become important soon.

We can easily estimate the convolution $h * \phi_y(t)$.

$$|h * \phi_y(t)| = \left|\int_Q \phi_y(t - x)\,h(x)\,dx\right|$$
$$= \left|\int_Q (\phi_y(t - x) - \phi_y(t - x_Q))\,h(x)\,dx\right|$$
$$\le C(d)(\ell(Q)/y)^\beta y^{-d}\|h\|_1$$
$$= C(d)\ell(Q)^\beta \|h\|_1 y^{-d-\beta},$$

implying that $A_\beta(h)(t, y) \le C(\beta, d)\ell(Q)^\beta\|h\|_1 y^{-d-\beta}$ for $(t, y) \in \Gamma(x)$. If we fix a $y > c'|x - x_Q|$ then

$$\int_{|t-x|<y} (A_\beta(f)(t, y))^2\,\frac{dt}{y^{d+1}} \le C(\beta, d)\ell(Q)^{2\beta}\|h\|_1^2 y^{-2d-2\beta-1}.$$

When we integrate this (in y) from $c'|x - x_Q|$ to infinity, and take a square root, we obtain:

$$G_\beta(h)(x) \le C(\beta, d)\ell(Q)^\beta\|h\|_1 |x - x_Q|^{-d-\beta}.$$

However, so long as $|x - x_Q| > \ell(Q)$,

$$
\frac{\ell(Q)^\beta}{|x - x_Q|^{d+\beta}} \leq C \frac{\ell(Q)^\beta}{(\ell(Q) + |x - x_Q|)^{d+\beta}}
$$
$$
= C\ell(Q)^{-d}(1 + |x - x_Q|/\ell(Q))^{-d-\beta}
$$
$$
= C|Q|^{-1}(1 + |x - x_Q|/\ell(Q))^{-d-\beta},
$$

which is exactly what we wanted. Theorem 6.1 is proved.

Knowing that $\|g_\psi(f)\|_p$ and $\|S_\psi(f)\|$ are both less than or equal to $C\|f\|_p$ for $1 < p < \infty$, we can use duality to prove the converse inequalities, much as we did in the dyadic setting (see exercise 3.7). If $f \in L^2 \cap L^p$ and $h \in L^2 \cap L^{p'}$, then (see equation 5.5),

$$
\int_{\mathbf{R}^d} f(x)\,\bar{h}(x)\,dx = \int_{\mathbf{R}^{d+1}_+} (f * \psi_y(t))\,(\bar{h} * \psi_y(t))\,\frac{dt\,dy}{y}.
$$

By the Cauchy-Schwarz inequality, the right-hand integral is seen to have modulus no bigger than

$$
\int_{\mathbf{R}^d} g_\psi(f)(x)\,g_\psi(h)(x)\,dx \leq \|g_\psi(f)\|_p \|g_\psi(h)\|_{p'}
$$
$$
\leq C\|g_\psi(f)\|_p \|h\|_{p'}.
$$

Therefore, by taking the supremum of $|\int f\,\bar{h}\,dx|$, as h runs over all of $L^2 \cap L^{p'}$, with $\|h\|_{p'} \leq 1$, we get $\|f\|_p \leq C\|g_\psi(f)\|_p$, for $f \in L^2 \cap L^p$. Now let f be an arbitrary element of L^p, and let f_n be a sequence in $L^2 \cap L^p$ such that $f_n \to f$ in L^p. We have the pointwise inequality $|g_\psi(f_n)(x) - g_\psi(f)(x)| \leq g_\psi(f_n - f)(x)$ (it's just the triangle inequality in disguise). But we also have that $\|g_\psi(f_n - f)\|_p \leq C\|f_n - f\|_p \to 0$. Therefore, $g_\psi(f_n) \to g_\psi(f)$ in L^p. Putting it all together:

$$
\|f\|_p = \lim_n \|f_n\|_p
$$
$$
\leq C \lim_n \|g_\psi(f_n)\|_p
$$
$$
= C\|g_\psi(f)\|_p,
$$

as was to be proved. The corresponding converse inequality for $S_\psi(f)$ is shown in the same fashion.

As often happens in analysis, we have proved—or almost proved—a little more than we advertised. The Cauchy-Schwarz inequality argument above implies that, if $f \in L^p$ and $h \in L^{p'}$, the integral

$$
\int_{\mathbf{R}^{d+1}_+} (f * \psi_y(t))\,(\bar{h} * \psi_y(t))\,\frac{dt\,dy}{y}
$$

is absolutely convergent, and the integral of the integrand's modulus is bounded by a constant times $\|f\|_p \|h\|_{p'}$. From this we can quickly get that

$$\int_{\mathbf{R}^d} f(x)\,\bar{h}(x)\,dx = \int_{\mathbf{R}^{d+1}_+} (f * \psi_y(t))\,(\bar{h} * \psi_y(t))\,\frac{dt\,dy}{y} \qquad (6.11)$$

for all $f \in L^p$ and $h \in L^{p'}$. Denote the integral on the right as $B(f,h)$. Let f_n be a sequence in $L^2 \cap L^p$ and let h_n be a sequence in $L^2 \cap L^{p'}$ such that $f_n \to f$ in L^p and $h_n \to h$ in $L^{p'}$. Then $\int f_n \bar{h}_n\,dx = B(f_n, h_n)$ for all n, and $\int f \bar{h}\,dx = \lim_n \int f_n \bar{h}_n\,dx$. However, mimicking the proof of the product rule, we have

$$|B(f_n, h_n) - B(f,h)| \leq C \left(\|f_n - f\|_p \|h_n\|_{p'} + \|f\|_p \|h_n - h\|_{p'} \right) \to 0,$$

implying $\int f \bar{h}\,dx = B(f,h)$.

We now have the equipment necessary to prove that the Calderón reproducing formula, in a very general sense, holds "in L^p" for all $1 < p < \infty$. (We will show that it holds in an even stronger sense somewhat later in this chapter.)

Theorem 6.2. *Let $f \in L^p$ ($1 < p < \infty$), and, for any measurable $K \subset \mathbf{R}^{d+1}_+$ with compact closure contained in $\mathbf{R}^{d+1}_+$, define, as we did in the previous chapter,*

$$f_{(K)}(x) = \int_K (f * \psi_y(t))\,\psi_y(x - t)\,\frac{dt\,dy}{y}.$$

If $K_1 \subset K_2 \subset \cdots$ is any compact-measurable exhaustion of $\mathbf{R}^{d+1}_+$, then $f_{(K_i)} \to f$ in L^p.

Proof of Theorem 6.2. For any measurable subset $A \subset \mathbf{R}^{d+1}_+$, temporarily define

$$S_A(f)(x) \equiv \left(\int_{\Gamma(x) \cap A} |f * \psi_y(t)|^2 \frac{dt\,dy}{y^{d+1}} \right)^{1/2}.$$

A repetition of the duality argument above shows that, if $K_m \subset K_n$, $\|f_{(K_m)} - f_{(K_n)}\|_p \leq C_p \|S_{K_n \setminus K_m}(f)\|_p$. But $S_{K_n \setminus K_m}(f)$ is pointwise dominated by $S_\psi(f)$, which belongs to L^p. Thus, $S_{K_n \setminus K_m}(f) \to 0$ almost everywhere and—by Dominated Convergence—also goes to 0 in L^p, as m and n go to infinity. Therefore the sequence $\{f_{(K_n)}\}_n$ is Cauchy in L^p. Denote its L^p limit by g. It will be enough to show that $f_{(K_n)} \to f$ weakly.

Let $h \in L^{p'}$. Then $\int f \bar{h}\,dx = B(f,h)$, where $B(\cdot, \cdot)$ is the bilinear functional we defined in 6.11. However, on the one hand, $B(f,h)$ is the limit of $B(f_{(K_n)}, h)$, by the Dominated Convergence Theorem; while, on the other hand, $B(f_{(K_n)}, h) \rightarrow B(g, h) = \int g\bar{h}\,dx$, because $f_{(K_n)} \rightarrow g$ in L^p. That finishes the proof.

Remark. It might seem that the proof of Theorem 6.2 makes those of Theorem 5.1 and Theorem 5.2 superfluous. Not quite. The proof of Theorem 6.2 rested on a reduction to L^2, which was handled by the two earlier theorems.

The square functions $S_\psi(f)$ and $g_\psi(f)$ are "semi-classical." The "classical" square functions $S_{cl}(f)$ and $g_{cl}(f)$ are defined as follows. Let $P : \mathbf{R}^d \mapsto \mathbf{R}$ be defined by

$$P(x) \equiv c_d(1 + |x|^2)^{(-d-1)/2},$$

where the constant c_d is chosen to ensure that $\int P(x)\,dx = 1$. If $f : \mathbf{R}^d \mapsto \mathbf{R}$ is such that $fP \in L^1$, then, for every $(x, y) \in \mathbf{R}_+^{d+1}$, we can define

$$u(x, y) \equiv P_y * f(x).$$

This function satisfies Laplace's equation $\Delta u \equiv 0$ in $\mathbf{R}_+^{d+1}$ and has boundary values equal to f, in the sense that

$$\lim_{y \to 0} u(x, y) = f(x)$$

almost everywhere, and $u(\cdot, y) \rightarrow f$ in L^p if $f \in L^p$ and $1 \leq p < \infty$. The gradient of u, ∇u, is a $(d+1)$-dimensional vector $\nabla u = (u_{x_1}, u_{x_1}, \ldots, u_{x_d}, u_y)$, all of whose components have essentially the same form. To wit, each u_{x_i} (or u_y) is given by a formula like

$$y^{-1}\phi_y * f(x),$$

where $\phi : \mathbf{R}^d \mapsto \mathbf{R}$ satisfies the following properties: 1) $|\phi(x)| \leq C(1+|x|)^{-d-1}$ for all x; 2) $|\nabla\phi(x)| \leq C(1 + |x|)^{-d-2}$ for all x; 3) $\int \phi(x)\,dx = 0$. In other words, the generating kernels for the components of ∇u are like non-compactly supported versions of the function ψ, in which the compact support condition has been replaced with uniform decay in the modulus and in the size of the derivatives.

The classical square functions are defined by the integrals:

$$S_{cl}(f)(x) = \left(\int_{\Gamma(x)} |\nabla u(t, y)|^2 \frac{dt\,dy}{y^{d-1}} \right)^{1/2}$$

$$g_{cl}(f) = \left(\int_0^\infty y|\nabla u(x, y)|^2\,dy \right)^{1/2}.$$

These are the ancestors of $S_\psi(f)$ and $g_\psi(f)$. They have a couple of advantages (but many disadvantages) when compared with their real-variable descendants. The main advantage is that, since they are defined via harmonic functions, tricks from classical harmonic analysis and partial differential equations (Green's Theorem, the Mean Value Property, etc.) can be used to control them. Unfortunately, this fact is usually swamped by one big disadvantage: they are generated by non-compactly supported kernels.

It turns out that $S_{cl}(f)$, $g_{cl}(f)$, and essentially all "reasonable" square functions are dominated by some $G_\beta(f)$, where β depends on the smoothness and rate of decay of the kernel used to define the square function. Proving this requires that we understand square functions that have the same form as $S_\psi(f)$ and $g_\psi(f)$, in which the function ψ is not assumed to have compact support, but to have "reasonable" decay in its size and smoothness (according to some measure), and of course to have integral equal to 0.

We will define two classes of such functions, each indexed over positive parameters β and ϵ, where we will (for obvious reasons) always assume that $\beta \leq 1$. The first class, $\mathcal{C}_{(\beta,\epsilon)}$, consists of functions that are only "pretty good," while those in the second class, $\mathcal{U}_{(\beta,\epsilon)}$, are "very good."

Definition 6.2. *We say that $\phi : \mathbf{R}^d \mapsto \mathbf{R}$ belongs to $\mathcal{C}_{(\beta,\epsilon)}$ if, for all $x \in \mathbf{R}^d$,*

$$|\phi(x)| \leq (1 + |x|)^{-d-\epsilon}; \tag{6.12}$$

and, for all x and x',

$$|\phi(x) - \phi(x')| \leq |x - x'|^\beta \left((1 + |x|)^{-d-\epsilon} + (1 + |x'|)^{-d-\epsilon} \right);$$

and if it also satisfies $\int \phi \, dx = 0$. We say that $\phi \in \mathcal{U}_{(\beta,\epsilon)}$ if ϕ satisfies 6.12, has integral equal to 0, and also, for all x and x',

$$|\phi(x) - \phi(x')| \leq |x - x'|^\beta \left((1 + |x|)^{-d-\epsilon-\beta} + (1 + |x'|)^{-d-\epsilon-\beta} \right). \tag{6.13}$$

The difference between the two classes is obvious: to belong to $\mathcal{U}_{(\beta,\epsilon)}$, we require that ϕ belong to $\mathcal{C}_{(\beta,\epsilon)}$ and have extra decay in its modulus of Hölder continuity, as expressed in 6.13. But this difference is something of an illusion. On the one hand, it is trivial that $\mathcal{U}_{(\beta,\epsilon)} \subset \mathcal{C}_{(\beta,\epsilon)}$. On the other hand, we have:

Lemma 6.2. *Let $0 < \beta \leq 1$ and $\epsilon > 0$, and suppose $0 < \beta' \leq \beta$ and $\beta' < \epsilon$. Define $\epsilon' \equiv \epsilon - \beta'$. Then:*

$$\mathcal{C}_{(\beta,\epsilon)} \subset \mathcal{U}_{(\beta',\epsilon')}.$$

Remark. We shall refer to Lemma 6.2 as the Free Lunch Lemma. The lemma means this: by sacrificing (actually, by not counting) a little of the decay of a $\phi \in \mathcal{C}_{(\beta,\epsilon)}$, we get a function (the *same* function!) with "improved" decay in its Hölder modulus, though to a different order.

Proof of the Free Lunch Lemma. Let $\phi \in \mathcal{C}_{(\beta,\epsilon)}$. It is trivial that

$$|\phi(x)| \leq (1 + |x|)^{-d-\epsilon'},$$

because $\epsilon' < \epsilon$. If $|x - x'| \leq 1$ then

$$|\phi(x) - \phi(x')| \le |x - x'|^\beta \left((1 + |x|)^{-d-\epsilon} + (1 + |x'|)^{-d-\epsilon} \right)$$
$$\le |x - x'|^{\beta'} \left((1 + |x|)^{-d-\epsilon'-\beta'} + (1 + |x'|)^{-d-\epsilon'-\beta'} \right),$$

because $|x - x'| \le 1$, $\beta' \le \beta$, and $\epsilon' + \beta' = \epsilon$. On the other hand, if $|x - x'| \ge 1$,

$$|\phi(x) - \phi(x')| \le (1 + |x|)^{-d-\epsilon} + (1 + |x'|)^{-d-\epsilon}$$
$$\le |x - x'|^{\beta'} \left((1 + |x|)^{-d-\epsilon'-\beta'} + (1 + |x'|)^{-d-\epsilon'-\beta'} \right).$$

The Free Lunch Lemma is proved.

For us, the usefulness of the Free Lunch Lemma comes from the following decomposition lemma due to A. Uchiyama [57]:

Lemma 6.3. *Let $0 < \beta \le 1$ and $\epsilon > 0$. There is a constant $C(d, \beta, \epsilon)$ such that, if $\psi \in \mathcal{U}_{(\beta,\epsilon)}$, there exists a sequence of functions $\{\phi_k\}_0^\infty$ such that*

$$\psi = C(d, \beta, \epsilon) \sum_0^\infty 2^{-k\epsilon} (\phi_k)_{2^k}$$

and every ϕ_k belongs to $\mathcal{C}_\beta$.

Proof of Lemma 6.3. Let $h \in \mathcal{C}_0^\infty(\mathbf{R}^d)$ be real, radial, non-negative, have support contained in $\{x : 1/4 \le |x| \le 1\}$, and be normalized so that

$$\sum_{-\infty}^\infty h(2^{-k}x) \equiv 1$$

for $x \ne 0$. There exists a $j_0 < 0$ such that

$$\sum_{j_0+1}^\infty h(2^{-k}x) \equiv 1$$

when $|x| \ge 1$. Re-index this sum as $\sum_1^\infty h(2^{-(j_0+k)}x)$. Define $\rho_0(x) \equiv 1 - \sum_{j_0+1}^\infty h(2^{-k}x)$ and set $\rho_k(x) \equiv h(2^{-(j_0+k)}x)$ for $k \ge 1$. Each ρ_k has support contained in $\{x : |x| \le 2^k\}$ and satisfies the inequalities:

$$\int \rho_k(x)\, dx \ge C2^{kd}$$
$$|\nabla \rho_k(x)| \le C2^{-k},$$

where the constants C only depend on d, h, and j_0.
Define, for $k \ge 0$,

$$g_k(x) \equiv \frac{\left(\int \left(\sum_0^k \rho_j \right) \psi\, dt \right) \rho_k(x)}{\int \rho_k(t)\, dt}.$$

It is easy to see that $g_k \to 0$ uniformly. In fact, this convergence is pretty fast. For any k,

$$\int \left(\sum_0^k \rho_j \right) \psi \, dt = - \int \left(\sum_{j>k} \rho_j \right) \psi \, dt, \tag{6.14}$$

because $\int \psi \, dt = 0$. But, when $k \geq 1$, the integral on the right-hand side of 6.14 has modulus no bigger than

$$\int_{|x| \geq c2^k} |\psi(t)| \, dt \leq C2^{-k\epsilon}.$$

Therefore, $g_k = C_k \rho_k$, where $|C_k| \leq C2^{-k(d+\epsilon)}$.

Notice that $\int g_0(t) \, dt = \int \psi \rho_0 \, dt$; and, when $k \geq 1$,

$$\int (g_k(t) - g_{k-1}(t)) \, dt = \int \psi(t) \rho_k(t) \, dt.$$

We can now finish the proof of Lemma 6.3 in a few lines. We write:

$$\psi = \sum_0^\infty \psi \rho_k = \sum_0^\infty \psi \rho_k - \left(g_0 + \sum_1^\infty (g_k - g_{k-1}) \right)$$

$$= (\psi \rho_0 - g_0) + \sum_1^\infty (\psi \rho_k - (g_k - g_{k-1}))$$

$$\equiv C(d, \beta, \epsilon) \sum_0^\infty 2^{-k\epsilon} \tilde{\phi}_k,$$

and we claim that, if we take $C(d, \beta, \epsilon)$ large enough, the ϕ_k's we have just (implicitly) defined will do the trick; i.e., that each $\tilde{\phi}_k$ equals $(\phi_k)_{2^k}$ for some $\phi_k \in \mathcal{C}_\beta$. This amounts to showing that, for each k, the support of $\tilde{\phi}_k$ is contained in $\{x : |x| \leq 2^k\}$, $\int \tilde{\phi}_k \, dt = 0$, and, for all x and x',

$$|\tilde{\phi}_k(x) - \tilde{\phi}_k(x')| \leq C|x - x'|^\beta 2^{-k(d+\beta)},$$

where C is a constant depending only on h, j_0, β, ϵ, and d. The first two requirements are easy; and the third is also not so bad. Without loss of generality we can assume that $k \geq 2$. Because of the support condition, we can assume that $|x - x'| \leq C2^k$; and, because each ρ_k is supported in an annulus, with inner and outer radii comparable to 2^k, we can assume that $|x|$ and $|x'|$ are both comparable to 2^k. For these x's and x''s we have the following estimates:

$$|\psi(x)| + |\psi(x')| \leq C2^{-k(d+\epsilon)}$$
$$|\psi(x) - \psi(x')| \leq C|x - x'|^\beta 2^{-k(d+\epsilon+\beta)}$$
$$|\rho_k(x)| + |\rho_k(x')| \leq C$$
$$|\rho_k(x) - \rho_k(x')| \leq C|x - x'|2^{-k} \leq C|x - x'|^\beta 2^{-k\beta},$$

where the last inequality uses the facts that $\beta \leq 1$ and $|x - x'|/2^k \leq C$. Combining these with our estimates on C_k above, we also get:

$$|g_k(x)| + |g_k(x')| \leq C2^{-k(d+\epsilon)}$$
$$|g_k(x) - g_k(x')| \leq C2^{-k(d+\epsilon+\beta)}|x - x'|^\beta,$$

which is exactly the smoothness we require. But now, by mimicking the proof of the product rule, it is easy to show that

$$|\psi(x)\rho_k(x) - \psi(x')\rho_k(x')| \leq C2^{-k(d+\epsilon+\beta)}|x - x'|^\beta,$$

which finishes the proof of Lemma 6.3.

Combined with the Free Lunch Lemma, Lemma 6.3 says that any $\phi \in C_{\beta,\epsilon}$ can be written, in a uniform manner, as a rapidly converging sum of dilates of functions in $C_{\beta'}$, for suitable $\beta' \leq \beta$.

We will be applying this fact shortly.

Definition 6.3. *Let β and ϵ be positive numbers, and suppose that $|f|(1 + |x|)^{-d-\epsilon} \in L^1$. For every $(t, y) \in \mathbf{R}_+^{d+1}$, set*

$$\tilde{A}_{(\beta,\epsilon)}(f)(t, y) \equiv \sup_{\phi \in C_{(\beta,\epsilon)}} |f * \phi_y(t)|.$$

Given this, the corresponding intrinsic square functions (non-compact support) are defined by:

$$\tilde{G}_{(\beta,\epsilon)}(f)(x) \equiv \left(\int_{\Gamma(x)} (\tilde{A}_{(\beta,\epsilon)}(f)(t, y))^2 \frac{dt\, dy}{y^{d+1}} \right)^{1/2}$$

$$\tilde{g}_{(\beta,\epsilon)}(f)(x) \equiv \left(\int_0^\infty (\tilde{A}_{(\beta,\epsilon)}(f)(x, y))^2 \frac{dy}{y} \right)^{1/2}$$

$$\tilde{\sigma}_{(\beta,\epsilon)}(f)(x) \equiv \left(\sum_{-\infty}^\infty (\tilde{A}_{(\beta,\epsilon)}(f)(x, 2^k))^2 \right)^{1/2}.$$

We have left it as an exercise to show that $\tilde{G}_{(\beta,\epsilon)}(f)$, $\tilde{g}_{(\beta,\epsilon)}(f)$, and $\tilde{\sigma}_{(\beta,\epsilon)}(f)$ are pointwise comparable, with comparability constants that only depend on d, β, and ϵ. Given this, the next theorem comes without too much extra work[4]:

Theorem 6.3. *Let $0 < \beta' \leq \beta$ and $\beta' < \epsilon$. There is a constant $C = C(\beta, \beta', \epsilon, d)$ such that, for all f having $|f|(1 + |x|)^{-d-\epsilon} \in L^1$,*

$$\tilde{G}_{(\beta,\epsilon)}(f) \leq CG_{\beta'}(f)$$

pointwise.

---

[4] Some work, but not too much.

Proof of Theorem 6.3. It's enough to show

$$\tilde{\sigma}_{(\beta,\epsilon)}(f)(0) \le C\sigma_{\beta'}(f)(0).$$

Set $\epsilon' = \epsilon - \beta'$. Let $\phi \in \mathcal{C}_{(\beta,\epsilon)}$. By the Free Lunch Lemma and Lemma 6.3, we can write

$$\phi = C(d, \beta', \epsilon) \sum_0^\infty 2^{-k\epsilon'}(\phi_k)_{2^k}$$

with every ϕ_k lying in $\mathcal{C}_{\beta'}$. Therefore,

$$|f * \phi(0)| \le C(d, \beta', \epsilon) \sum_0^\infty 2^{-k\epsilon'} A_{\beta'}(f)(0, 2^k)$$

$$\le C(d, \beta', \epsilon) \left(\sum_0^\infty 2^{-k\epsilon'} (A_{\beta'}(f)(0, 2^k))^2 \right)^{1/2}.$$

If we dilate both sides of the preceding inequality by 2^j, we get, for each j,

$$|f * \phi_{2^j}(0)| \le C(d, \beta', \epsilon) \left(\sum_0^\infty 2^{-k\epsilon'} (A_{\beta'}(f)(0, 2^{k+j}))^2 \right)^{1/2}.$$

Now let $\{\phi^{(j)}\}_{-\infty}^\infty$ be an arbitrary sequence of functions from $\mathcal{C}_{(\beta,\epsilon)}$. Summing the preceding inequality over j yields:

$$\sum_{j=-\infty}^\infty |f * \phi_{2^j}^{(j)}(0)|^2 \le C(d, \beta', \epsilon) \sum_{j=-\infty}^\infty \sum_{k=0}^\infty 2^{-k\epsilon'} (A_{\beta'}(f)(0, 2^{k+j}))^2$$

$$= C(d, \beta', \epsilon) \sum_{l=-\infty}^\infty (A_{\beta'}(f)(0, 2^l))^2 \sum_{j=-\infty}^l 2^{-(l-j)\epsilon'}$$

$$= C(d, \beta', \epsilon) \sum_{l=-\infty}^\infty (A_{\beta'}(f)(0, 2^l))^2 \sum_{k=0}^\infty 2^{-k\epsilon'}$$

$$= C(d, \beta', \epsilon) \sum_{l=-\infty}^\infty (A_{\beta'}(f)(0, 2^l))^2$$

$$= C(d, \beta', \epsilon)(\sigma_{\beta'}(f)(0))^2;$$

which is what we want, because the supremum of the left-hand side—over all such sequences $\{\phi^{(j)}\}_{-\infty}^\infty$—is $(\tilde{\sigma}_{(\beta,\epsilon)}(f)(0))^2$. Theorem 6.3 is proved.

Since the kernels that generate the components of ∇u all belong to $\mathcal{U}_{(1,1)}$, the inequality,

$$S_{cl}(f) + g_{cl}(f) \le CG_1(f),$$

is now trivial, implying that $\|S_{cl}(f)\|_p + \|g_{cl}(f)\|_p \le C\|f\|_p$ for all $1 < p < \infty$. As with the ψ-defined square functions we saw earlier, these classical square functions also satisfy converse inequalities, and these converse inequalities are also proved via duality. This duality argument will go through more smoothly if we first prove a very general version of the Calderón reproducing formula.

Let ϕ and ψ be two positive multiples of functions in $\mathcal{C}_{(\beta,\epsilon)}$. We also require that ϕ and ψ be real, radial, and "co-normalized" so that

$$\int_0^\infty \hat{\phi}(y\xi)\, \hat{\psi}(y\xi)\, \frac{dy}{y} = 1 \qquad (6.15)$$

for all $\xi \neq 0$. (It is this co-normalization that makes us resort to *positive multiples* of function in $\mathcal{C}_{(\beta,\epsilon)}$.) For now we will also require that the integral 6.15 be absolutely convergent. In an exercise it is shown that this is no restriction.

Theorem 6.4. . *Let ϕ and ψ be as in the preceding paragraph. Let f be such that $|f|(1+|x|)^{-d-\epsilon} \in L^1$. For any measurable K with compact closure contained in $\mathbf{R}_+^{d+1}$, define*

$$f_{(K),\phi,\psi}(x) = \int_K (f * \phi_y(t))\, \psi_y(x-t)\, \frac{dt\, dy}{y}. \qquad (6.16)$$

If $f \in L^p$ ($1 < p < \infty$) and $K_1 \subset K_2 \subset \cdots$ is any compact-measurable exhaustion of $\mathbf{R}_+^{d+1}$, then $f_{(K_i),\phi,\psi} \to f$ in L^p.

Proof of Theorem 6.4. If $f \in L^2$, then

$$\int |f(x)|^2\, dx = \int |\hat{f}(\xi)|^2\, d\xi$$

$$= \int |\hat{f}(\xi)|^2 \left(\int_0^\infty \hat{\phi}(y\xi)\, \hat{\psi}(y\xi)\, \frac{dy}{y} \right) d\xi$$

$$= \int_0^\infty \left(\int \hat{f}(\xi)\, \hat{\phi}(y\xi)\, \overline{\hat{f}(\xi)}\, \hat{\psi}(y\xi)\, d\xi \right) \frac{dy}{y}$$

$$= \int_{\mathbf{R}_+^{d+1}} (f * \phi_y(t))\, (\bar{f} * \psi_y(t))\, \frac{dt\, dy}{y},$$

where every equation follows from Plancherel's Theorem and the fact that the integrals are all absolutely convergent. (To bound the last one, note that

$$\int_{\mathbf{R}_+^{d+1}} |f * \phi_y(t)||\bar{f} * \psi_y(t)|\, \frac{dt\, dy}{y} \leq C \int (\tilde{g}_{(\beta,\epsilon)}(f)(t))^2\, dt.)$$

Polarization shows that, for every $h \in L^2$,

$$\int_{\mathbf{R}^d} f(x)\, \bar{h}(x)\, dx = \int_{\mathbf{R}_+^{d+1}} (f * \phi_y(t))\, (\bar{h} * \psi_y(t))\, \frac{dt\, dy}{y}. \qquad (6.17)$$

Both integrals are absolutely convergent. Indeed, the second integral is bounded by

$$\int_{\mathbf{R}^d} \tilde{g}_{(\beta,\epsilon)}(f)(t)\, \tilde{g}_{(\beta,\epsilon)}(h)(t)\, dt.$$

Because of this absolute convergence, the integral on the right-hand side of 6.17 is equal to the limit of

$$\int f_{(K_i),\phi,\psi}(x)\, \bar{h}(x)\, dx = \int_{K_i} (f * \phi_y(t))\, (\bar{h} * \psi_y(t))\, \frac{dt\, dy}{y}.$$

Therefore, if $f \in L^2$, the sequence $f_{(K_i),\phi,\psi} \to f$ weakly in L^2. To prove that $f_{(K_i)} \to f$ in L^p when $f \in L^p$, we only need to show that, for every $f \in L^p$, the sequence $\{f_{(K_i),\phi,\psi}\}$ is Cauchy in L^p.

By considering integrals of the form $\int f_{(K_i),\phi,\psi}\, \bar{h}\, dx$ for arbitrary $f \in L^2 \cap L^p$ and $h \in L^2 \cap L^{p'}$, we see that

$$\left| \int f_{(K_i),\phi,\psi}(x)\, \bar{h}(x)\, dx \right| \leq \int \tilde{g}_{(\beta,\epsilon)}(f)(x)\, \tilde{g}_{(\beta,\epsilon)}(h)(x)\, dx$$

$$\leq C \|\tilde{g}_{(\beta,\epsilon)}(f)\|_p \|\tilde{g}_{(\beta,\epsilon)}(h)\|_{p'},$$

implying that $\|f_{(K_i),\phi,\psi}\|_p \leq C\|\tilde{g}_{(\beta,\epsilon)}(f)\|_p$. But more is true. For $E \subset \mathbf{R}_+^{d+1}$, define

$$\tilde{g}_{(\beta,\epsilon,E)}(f)(x) = \left(\int_{y:\ (x,y)\in E} (\tilde{A}_{(\beta,\epsilon)}(f)(x,y))^2\, \frac{dy}{y} \right)^{1/2}.$$

The duality argument shows that

$$\|f_{(K_i),\phi,\psi}\|_p \leq C\|\tilde{g}_{(\beta,\epsilon,K_i)}(f)\|_p, \tag{6.18}$$

with a constant independent of K_i, for all $f \in L^2 \cap L^p$. But, because K_i has compact closure contained in $\mathbf{R}_+^{d+1}$, it is trivial that both sides of 6.18 vary continuously with $f \in L^p$, relative to the L^p norm. Therefore 6.18 holds for *all* $f \in L^p$.

Continuing on this line, if $f \in L^p$ and $K_m \subset K_n$, then

$$\|f_{(K_m),\phi,\psi} - f_{(K_n),\phi,\psi}\|_p \leq C\|\tilde{g}_{(\beta,\epsilon,K_n\setminus K_m)}(f)\|_p.$$

A (by now) familiar argument shows that

$$\|\tilde{g}_{(\beta,\epsilon,K_n\setminus K_m)}(f)\|_p \to 0$$

as m and n go to infinity. Therefore $\{f_{(K_n),\phi,\psi}\}$ is Cauchy in L^p. This finishes the proof of Theorem 6.4.

It is now easy to show that $\|f\|_p \leq C_p\|g_{cl}(f)\|_p$. In fact, we've essentially just shown it.

We will prove the inequality for something *smaller* than $g_{cl}(f)$; namely,

$$g_0(f)(x) \equiv \left(\int_0^\infty y \left| \frac{\partial u}{\partial y}(x,y) \right|^2 dy \right)^{1/2}.$$

The partial derivative $\frac{\partial u}{\partial y}$ is given by the formula

$$\frac{\partial u}{\partial y}(x,y) = y^{-1} \phi_y * f(x),$$

where ϕ is real, radial, and, modulo a scalar multiple, belongs to $\mathcal{U}_{(1,1)}$. We can find a real, radial $\psi \in \mathcal{C}_0^\infty(\mathbf{R}^d)$ with integral equal to 0, and normalized so that

$$\int_0^\infty \hat{\phi}(\xi y)\, \hat{\psi}(\xi y)\, \frac{dy}{y} \equiv 1$$

for all $\xi \neq 0$. By the Calderón reproducing formula method, we can write

$$f = \int_{\mathbf{R}_+^{d+1}} (f * \phi_y(t))\, \psi_y(x-t)\, \frac{dt\,dy}{y},$$

in the sense that f equals the L^p limit of integrals on the right, taken over a compact-measurable exhaustion of $\mathbf{R}_+^{d+1}$. However, this is the same as saying

$$f = \int_{\mathbf{R}_+^{d+1}} \frac{\partial u}{\partial y}(t,y)\, \psi_y(x-t)\, dt\,dy.$$

For $K \subset \mathbf{R}_+^{d+1}$ a compact set, define

$$f_{(K)} = \int_K \frac{\partial u}{\partial y}(t,y)\, \psi_y(x-t)\, dt\,dy.$$

Choose K so that $\|f - f_K\|_p < \epsilon$, and let $h \in L^{p'}$ have unit norm. Then:

$$\left| \int_{\mathbf{R}^d} f(x)\, \bar{h}(x)\, dx \right| \leq \epsilon + \left| \int_K \frac{\partial u}{\partial y}(t,y)\, (\bar{h} * \psi_y(t))\, dt\,dy \right|$$

$$\leq \epsilon + \int_{\mathbf{R}^d} g_0(f)(t)\, g_\psi(h)(t)\, dt$$

$$\leq \epsilon + C \|g_0(f)\|_p,$$

which proves $\|f\|_p \leq C \|g_0(f)\|_p$.

There is clearly nothing terribly special about the kernel defining g_0. The preceding argument actually gives us the following.

Theorem 6.5. *Let $\phi \in \mathcal{C}_{(\beta,\epsilon)}$ be real and radial, and suppose that $\phi \not\equiv 0$. If $1 < p < \infty$, then, for all $f \in L^p$,*

$$\|f\|_p \sim \|S_\phi(f)\|_p \sim \|g_\phi(f)\|_p,$$

with comparability constants depending only on p, ϕ, and d; and where we are defining

$$S_\phi(f)(x) \equiv \left(\int_{\Gamma(x)} |f * \phi_y(t)|^2 \frac{dt\,dy}{y^{d+1}} \right)^{1/2}$$

$$g_\phi(f)(x) \equiv \left(\int_0^\infty |f * \phi_y(x)|^2 \frac{dy}{y} \right)^{1/2}.$$

Proof of Theorem 6.5. Our theorems on the intrinsic square function show that the L^p norms of $S_\phi(f)$ and $g_\phi(f)$ are dominated by $\|f\|_p$. For the other direction, note that, up to multiplication by a positive constant, the pair of functions (ϕ, ϕ) satisfies 6.15. The result now follows by a virtual repetition of the argument we just gave for $g_0(f)$.

Exercises

6.1. For $\beta > 0$ and $(t, y) \in \mathbf{R}_+^{d+1}$, let $\mathcal{C}_\beta(t, y)$ be the family of functions $\phi : B(t; y) \mapsto \mathbf{R}$ such that $\int \phi\, dx = 0$ and, for all x and x', $|\phi(x) - \phi(x')| \leq y^{-d-\beta}|x - x'|^\beta$. Show that

$$A_\beta(f)(t, y) = \sup_{\phi \in \mathcal{C}_\beta(t,y)} \left| \int f(x)\,\phi(x)\,dx \right|.$$

6.2. Show that, if $\alpha \geq 1$, then $G_{\alpha,\beta}(f) \leq C(\beta, d)\alpha^M G_\beta(f)$, where M is an exponent depending on β and d. Also, find the value of M. (Hint: Exercise 6.1 will come in useful here.)

6.3. Prove that $G_\beta(f) \sim g_\beta(f) \sim \sigma_\beta(f)$ pointwise, with comparability constants that only depend on β and d. Exercise 6.1 will be useful here too.

6.4. Prove that $\tilde{G}_{(\beta,\epsilon)}(f)$, $\tilde{g}_{(\beta,\epsilon)}(f)$, and $\tilde{\sigma}_{(\beta,\epsilon)}(f)$ are pointwise comparable, with comparability constants that only depend on d, β, and ϵ.

6.5. Theorem 6.1 was proved with the restriction that f's averages over large cubes got very small. To remove it, we consider the modified square function $G_{\beta,K}(f)$,

$$G_{\beta,K}(f)(x) \equiv \left(\int_{\Gamma(x) \cap K} (A_\beta(f)(t, y))^2 \frac{dt\,dy}{y^{d+1}} \right)^{1/2},$$

where K is a measurable set with compact closure contained in $\mathbf{R}_+^{d+1}$. Theorem 6.1 will follow if we can prove it for $G_{\beta,K}(f)$, with constants independent of K. Show that this approach will solve our problem. The main thing to observe here is that, for every such K, there is an $R > 0$ such that, for all f, $G_{\beta,K}(f) = G_{\beta,K}(f\chi_{B(0;R)})$, and the function $f\chi_{B(0;R)}$ has the "small average" property if $f \in L^1_{\text{loc}}(\mathbf{R}^d)$.

6.6. The proof of Lemma 6.3 calls for a function $h \in \mathcal{C}_0^\infty(\mathbf{R}^d)$ that is real, radial, non-negative, has support contained in $\{x : 1/4 \leq |x| \leq 1\}$, and satisfies

$$\sum_{-\infty}^{\infty} h(2^{-k}x) \equiv 1$$

for $x \neq 0$. Show that such a function exists. Also, show that, for such an h, there is a $j_0 < 0$ such that

$$\sum_{j_0+1}^{\infty} h(2^{-k}x) \equiv 1$$

when $|x| \geq 1$.

6.7. Let $\phi \in \mathcal{U}_{(\beta,\epsilon)}$. Show that there are positive numbers δ_1 and δ_2, and a positive constant C, depending only on β, ϵ, and d, so that, for all ξ:

$$|\hat{\phi}(\xi)| \leq C \min(|\xi|^{\delta_1}, |\xi|^{-\delta_2}).$$

Then show how this implies that the integral 6.15 is automatically absolutely convergent for ϕ and ψ in $\mathcal{C}_{(\beta,\epsilon)}$.

6.8. Let $f \in L^p$ $(1 < p < \infty)$, and suppose that ϕ and ψ are two real, radial scalar multiples of functions in $\mathcal{C}_{(\beta,\epsilon)}$ satisfying 6.15. We know that

$$f(x) = \int_{\mathbf{R}_+^{d+1}} (f * \phi_y(t))\, \psi_y(x - t)\, \frac{dt\, dy}{y}$$

as an L^p limit over any compact-measurable exhaustion $\{K_i\}$ of $\mathbf{R}_+^{d+1}$. Show that, if $E \subset \mathbf{R}_+^{d+1}$ is measurable, then

$$f_{(E),\phi,\psi}(x) \equiv \lim \int_{E \cap K_i} (f * \phi_y(t))\, \psi_y(x - t)\, \frac{dt\, dy}{y}$$

exists as an L^p limit, independent of the sequence $\{K_i\}$. Show that this definition of $f_{(E),\phi,\psi}$ coincides with our earlier formula 6.16 when E has compact closure contained in $\mathbf{R}_+^{d+1}$. Show that, if F and E are arbitrary measurable subsets of $\mathbf{R}_+^{d+1}$, and $E \subset F$, then $f_{(F)} - f_{(E)} = f_{(F \setminus E)}$. Show that, if $\{E_i\}$ is any increasing sequence of measurable sets such that $\cup_i E_i = \mathbf{R}_+^{d+1}$, then $f_{(E_i),\phi,\psi} \to f$ in L^p.

6.9. Suppose we have μ, a positive Borel measure on $\mathbf{R}_+^{d+1}$, and that μ satisfies

$$\mu(T(Q)) \leq |Q|$$

for all cubes $Q \subset \mathbf{R}^d$. Let β and ϵ be two positive numbers, and let ϕ and ψ be two fixed functions in $\mathcal{C}_{(\beta,\epsilon)}$. We want to make sense of the following operator:

$$T(f)(x) \equiv \int_{\mathbf{R}_+^{d+1}} (f * \psi_y(t)) \, \phi_y(x - t) \, d\mu(t, y). \qquad (6.19)$$

a) Show that, if $1 < p < \infty$, $f \in L^p$, and $\{K_i\}_i$ is any compact-(Borel)measurable exhaustion of $\mathbf{R}_+^{d+1}$, then the sequence of functions

$$\int_{K_i} (f * \psi_y(t)) \, \phi_y(x - t) \, d\mu(t, y)$$

has an L^p limit that is independent of the collection $\{K_i\}_i$. In the process you will show that this limit has L^p norm less than or equal to a constant times $\|f\|_p$, with a constant that only depends on β, ϵ, p, and d. "This limit," of course, is what we formally mean by 6.19.

b) Now we are going to use part a) to show that the Calderón reproducing formula has a fair amount of redundancy built into it. Let $\psi \in \mathcal{C}_0^\infty(\mathbf{R}^d)$ be real, radial, have support contained in $\{x : |x| \leq 1\}$, and satisfy $\int \psi \, dx = 0$ and 5.1. Let $E \subset \mathbf{R}_+^{d+1}$ be a measurable set such that, for some $\delta > 0$,

$$|E \cap T(Q)| \leq \delta |T(Q)|$$

for all cubes $Q \subset \mathbf{R}^d$. Let T_E be the operator defined by

$$T_E(f)(x) \equiv \int_E (f * \psi_y(t)) \, \psi_y(x - t) \, \frac{dt \, dy}{y},$$

where the integral is defined as in problem 6.8. Use part a) to show that $\|T_E(f)\|_p \leq C\delta \|f\|_p$, where C depends on p, d, and ψ, but not on δ. Use this to show that, if δ is small enough, the operator $I - T_E$ (where I is the identity) is *invertible* on L^p. In other words, if E is sparse enough on all of the $T(Q)$'s, and we throw out the values of $f * \psi_y(t)$ on E, we can still recover all of f from the remaining values.

6.10. Suppose we have a family of functions $\{\phi_{(Q)}\}_Q$, indexed over the family of dyadic cubes $\mathcal{D}_d$. We also suppose that, for some fixed positive numbers α and ϵ, each $\phi_{(Q)}$ satisfies

$$|\phi_{(Q)}(x)| \leq |Q|^{-1/2} \left(1 + \frac{|x - x_Q|}{\ell(Q)}\right)^{-d-\epsilon}$$

$$|\phi_{(Q)}(x) - \phi_{(Q)}(x')| \leq \left(\frac{|x - x'|}{\ell(Q)}\right)^\alpha |Q|^{-1/2}$$

$$\times \left(\left(1 + \frac{|x - x_Q|}{\ell(Q)}\right)^{-d-\epsilon} + \left(1 + \frac{|x' - x_Q|}{\ell(Q)}\right)^{-d-\epsilon}\right)$$

for all x and x' in $\mathbf{R}^d$ (where we are using x_Q to denote Q's center); and furthermore

$$\int \phi_{(Q)}(x)\, dx = 0.$$

Show that the family $\{\phi_{(Q)}\}_Q$ is "almost-orthogonal" in the following sense: there is a constant C, depending only on α, ϵ, and d, such that, for all finite linear sums $\sum \lambda_Q \phi_{(Q)}$,

$$\int \Big| \sum \lambda_Q \phi_{(Q)} \Big|^2 \leq C \sum |\lambda_Q|^2.$$

Then state and prove a generalization of this result to $L^p(\mathbf{R}^d)$, for $1 < p < \infty$. (Hint: Begin by showing that, if $g \in L^2(\mathbf{R}^d)$, then

$$\left| \int \Big(\sum \lambda_Q \phi_{(Q)} \Big) g(x)\, dx \right| \leq C \int H(x)\, G(g)(x)\, dx,$$

where $G(g)$ is the intrinsic square function of g [to some order, depending on α and ϵ] and

$$H(x) = \left(\sum_Q \frac{|\lambda_Q|^2}{|Q|} \chi_Q(x) \right)^{1/2}.$$

6.11. In our proof of Theorem 6.4, we wrote: "Therefore, if $f \in L^2$, the sequence $f_{(K_i),\phi,\psi} \to f$ weakly in L^2. To prove that $f_{(K_i)} \to f$ in L^p, when $f \in L^p$, we only need to show that, for every $f \in L^p$, the sequence $\{f_{(K_i),\phi,\psi}\}$ is Cauchy in L^p." There is a small logical gap between the first and second sentences. Find it and fill it.

6.12. Suppose that ϕ and ψ are two positive multiples of functions in $\mathcal{C}_{(\beta,\epsilon)}$, satisfying 6.15, and $1 < p < \infty$. Show that, if $f \in L^p$ and $h \in L^{p'}$, then

$$\int_{\mathbf{R}^d} f(x)\, \bar{h}(x)\, dx = \int_{\mathbf{R}^{d+1}_+} (f * \phi_y(t))\, (\bar{h} * \psi_y(t)) \, \frac{dt\, dy}{y},$$

and that the integral on the right is absolutely convergent.

6.13. We justified inequality 6.6 by referring to "our earlier work" on adapted functions, which involved some roundabout manipulations (reductions to the dyadic case, good-λ inequalities, etc.). Here is a more direct way to get 6.6. First, assume that all the cube triples $\tilde{Q}$ under consideration belong to a good family $\mathcal{G}$. Then, by a familiar reduction, assume that every function $a_{(Q)}$ is adapted to a dyadic cube Q. Next show that, for all Q and Q' in $\mathcal{D}_d$,

$$\left| \int a_{(Q)}(x)\, a_{(Q')}(x)\, dx \right| \leq \begin{cases} 0 & \text{if } Q \cap Q' = \emptyset; \\ C(|Q|^{1/2}/|Q'|^{1/2})(\ell(Q)^\beta/\ell(Q')^\beta) & \text{if } Q \subset Q'. \end{cases}$$

(See Lemma 4.1.) Therefore, for any finite sum $\sum \gamma_Q a_{(Q)}$, where the $a_{(Q)}$'s satisfy 1., 2., and, 3.,

$$\int |\sum_Q \gamma_Q a_{(Q)}|^2 dx \leq C \sum_Q |\gamma_Q| \left(\sum_{Q':Q\subset Q'} |\gamma_{Q'}| \left(\frac{|Q|^{1/2}}{|Q'|^{1/2}} \right) \left(\frac{\ell(Q)^\beta}{\ell(Q')^\beta} \right) \right).$$
(6.20)

For each fixed Q,

$$\sum_{Q':Q\subset Q'} |\gamma_{Q'}| \left(\frac{|Q|^{1/2}}{|Q'|^{1/2}} \right) \left(\frac{\ell(Q)^\beta}{\ell(Q')^\beta} \right) \leq$$
$$\left(\sum_{Q':Q\subset Q'} |\gamma_{Q'}|^2 \left(\frac{|Q|}{|Q'|} \right) \frac{\ell(Q)^\beta}{\ell(Q')^\beta} \right)^{1/2} \left(\sum_{Q':Q\subset Q'} \frac{\ell(Q)^\beta}{\ell(Q')^\beta} \right)^{1/2}$$

by Cauchy-Schwarz. But

$$\sum_{Q':Q\subset Q'} \frac{\ell(Q)^\beta}{\ell(Q')^\beta} \leq C_\beta.$$

Therefore the left-hand side of 6.20 is less than or equal to a constant times

$$\sum_Q |\gamma_Q| \left(\sum_{Q':Q\subset Q'} |\gamma_{Q'}|^2 \left(\frac{|Q|}{|Q'|} \right) \left(\frac{\ell(Q)^\beta}{\ell(Q')^\beta} \right) \right)^{1/2};$$

which (again by Cauchy-Schwarz) is less than or equal to

$$\left(\sum_Q |\gamma_Q|^2 \right)^{1/2} \left(\sum_{Q'} |\gamma_{Q'}|^2 \sum_{Q:Q\subset Q'} \left(\frac{|Q|}{|Q'|} \right) \left(\frac{\ell(Q)^\beta}{\ell(Q')^\beta} \right) \right)^{1/2}.$$

The first factor (containing the single sum, over Q) is what we want. The second factor (containing the iterated sum) is also okay; because, for every fixed $Q' \in \mathcal{D}_d$,

$$\sum_{Q:Q\subset Q'} \left(\frac{|Q|}{|Q'|} \right) \left(\frac{\ell(Q)^\beta}{\ell(Q')^\beta} \right) \leq C_\beta.$$
(6.21)

Proving 6.21 will give the reader valuable practice with the geometry of dyadic cubes.

Notes

The intrinsic square function first appears in [59]. An early form of $g_{cl}(f)$ first appeared in the papers [37] [38] [39] of Littlewood and Paley. The classical theory (based on analytic and harmonic functions) was further developed in [66], [51], and [52]. An excellent classical treatment of $S_{cl}(f)$ and $g_{cl}(f)$ is in Chapter IV of [53], to which the approach taken here is indebted. We refer the reader to Stein's article [54] for a historical overview of classical Littlewood-Paley theory. To the author's best knowledge, the earliest appearance of a real-variable square function having the form of $S_{\psi,\alpha}(f)$ is in [9].

7

The Calderón Reproducing Formula III

We have seen that, if $1 < p < \infty$, the Calderón reproducing formula converges "in $L^p(\mathbf{R}^d, dx)$" under very general hypotheses. In this chapter we will see that it converges, in the same sense, in $L^p(\mathbf{R}^d, w\,dx) \equiv L^p(w)$, for a particular class of weights, called A_p. In the process we will have to prove some facts about A_p, and this process should help to explain why this class is the natural one for which such a result should hold. It is also natural for a lot of other things in analysis. Our arguments will (we hope) show this too.

Recall that, if $1 < p < \infty$, we set $p' = p/(p-1)$, the dual exponent to p.

We say that a weight w belongs to the Muckenhoupt class A_p ($1 < p < \infty$) if

$$\sup_{Q \subset \mathbf{R}^d} \left(\frac{1}{|Q|} \int_Q w(x)\,dx \right)^{1/p} \left(\frac{1}{|Q|} \int_Q w(x)^{1-p'}\,dx \right)^{1/p'} < \infty, \qquad (7.1)$$

where the supremum is over all bounded cubes $Q \subset \mathbf{R}^d$. The value of the supremum is called w's A_p "norm" and is denoted by $\|w\|_{A_p}$.

The first things to observe about A_p are that $w \in A_p$ if and only if $w^{1-p'} \in A_{p'}$, and $\|w\|_{A_p} = \|w^{1-p'}\|_{A_{p'}}$. The reason for these is that, if we replace w by $w^{1-p'}$ and p by p' in 7.1, we get an *identical* condition. Proving this will give the reader valuable practice with manipulating the relationships between p and p', which include:

$$pp' = p + p';$$
$$(1-p)(1-p') = 1;$$
$$-p'/p = 1 - p'.$$

We are going to show that the Calderón reproducing formula converges in $L^p(w)$ ($1 < p < \infty$) for all A_p weights w. The formal statement of this is the following theorem.

Theorem 7.1. *Let β and ϵ be positive numbers. Suppose that ϕ and ψ are real and radial, and are positive multiples of functions in $\mathcal{C}_{(\beta,\epsilon)}$. We furthermore assume that ϕ and ψ are "co-normalized" so that*

$$\int_0^\infty \hat{\phi}(y\xi)\,\hat{\psi}(y\xi)\,\frac{dy}{y} = 1$$

for all $\xi \neq 0$. Let $1 < p < \infty$ and suppose $w \in A_p$. For every $f \in L^p(w)$ and measurable K with compact closure contained in $\mathbf{R}_+^{d+1}$, set

$$f_{(K),\phi,\psi}(x) = \int_K (f * \phi_y(t))\,\psi_y(x - t)\,\frac{dt\,dy}{y}.$$

If $K_1 \subset K_2 \subset \cdots$ is any compact-measurable exhaustion of $\mathbf{R}_+^{d+1}$, then $f_{(K_i),\phi,\psi} \to f$ in $L^p(w)$.

Theorem 7.1 will follow almost immediately from

Theorem 7.2. *Let $0 < \beta \leq 1$ and $1 < p < \infty$. If $w \in A_p$, there is a constant $C = C(\beta, w, p, d)$ such that, for all $f \in L^p(w)$,*

$$\int_{\mathbf{R}^d} (G_\beta(f)(x))^p\,w\,dx \leq C \int_{\mathbf{R}^d} |f(x)|^p\,w\,dx. \tag{7.2}$$

Here is how Theorem 7.2 will *almost* give us Theorem 7.1. Let $f \in L^p(w)$ and let g be any bounded, compactly supported function defined on $\mathbf{R}^d$. Set $h \equiv gw$. Let $K \subset \mathbf{R}^d$ be as in the hypotheses of Theorem 7.1. We have made it an exercise (see 7.2; it has lots of hints) to show that $f * \phi_y(t)$ is defined and bounded on K (this is the "almost" part at the beginning of the paragraph). Assuming that fact, we have:

$$\int_{\mathbf{R}^d} f_{(K_i),\phi,\psi}(x)\,g(x)\,w\,dx = \int_K (f * \phi_y(t))\,(h * \psi_y(t))\,\frac{dt\,dy}{y}. \tag{7.3}$$

Define

$$g_\phi(f)(x) = \left(\int_0^\infty |f * \phi_y(x)|^2\,\frac{dy}{y} \right)^{1/2}$$

$$g_{\phi,K_i}(f)(x) = \left(\int_{\{y:\ (x,y)\in K_i\}} |f * \phi_y(x)|^2\,\frac{dy}{y} \right)^{1/2}$$

$$g_\psi(h)(x) = \left(\int_0^\infty |h * \psi_y(x)|^2\,\frac{dy}{y} \right)^{1/2}.$$

By Cauchy-Schwarz, the absolute value of the right-hand side of 7.3 is bounded by

$$\int_{\mathbf{R}^d} (g_{\phi,K_i}(f)(x))\,(g_\psi(h)(x))\,dx,$$

which we will write as

$$\int_{\mathbf{R}^d} \left[(g_{\phi,K_i}(f)(x))w^{1/p} \right] \left[(g_\psi(h)(x))w^{-1/p} \right] dx.$$

If we have Theorem 7.2, this integral will be bounded by a constant times the product of $\|(g_{\phi,K_i}(f)\|_{L^p(w)}$ and

$$\left(\int_{\mathbf{R}^d} (G_{\tilde\beta}(h)(x))^{p'} w^{-p'/p} dx \right)^{1/p'}, \tag{7.4}$$

for some $0 < \tilde\beta \le \beta$. However, $w^{-p'/p} = w^{1-p'} \in A_{p'}$. Applying Theorem 7.2 again lets us dominate 7.4 by a constant times

$$\left(\int_{\mathbf{R}^d} |h(x)|^{p'} w^{1-p'} dx \right)^{1/p'};$$

which, after unraveling the exponents, turns out to be $\|g\|_{L^{p'}(w)}$. Taking the supremum over all such g satisfying $\|g\|_{L^{p'}(w)} \le 1$, we get

$$\|f_{(K_i),\phi,\psi}\|_{L^p(w)} \le C\|g_{\phi,K_i}(f)\|_{L^p(w)}.$$

But we also have that

$$\|g_{\phi,K_i}(f)\|_{L^p(w)} \le C\|(G_{\tilde\beta}(f)\|_{L^p(w)} \le C\|f\|_{L^p(w)} < \infty,$$

implying that $g_\phi(f) < \infty$ w-almost-everywhere. Therefore (arguing exactly as we did in the unweighted setting), if $\{K_i\}$ is any compact-measurable exhaustion, the sequence $\{f_{(K_i),\phi,\psi}\}$ will be Cauchy in $L^p(w)$. It is not hard (we leave it as an exercise) to show that this limit will be independent of the exhaustion $\{K_i\}$. If f is bounded and has bounded support, the limit of the $f_{(K_i),\phi,\psi}$'s must be f (because we can always take a subsequence converging to f almost everywhere). Therefore the limit must be f for *all* $f \in L^p(w)$.

So, everything comes down to showing Theorem 7.2 and the boundedness of $f * \phi_y(t)$ on K. For now we will concentrate on Theorem 7.2. It will follow from some work on A_p weights and:

Theorem 7.3. *Let X be a non-empty set, and let T_1 and T_2 be two operators mapping from X into $L^1_{\mathrm{loc}}(\mathbf{R}^d)$. Suppose there is a number $p_0 > 1$ such that, for all $1 < q < p_0$, all weights v, and all $f \in X$,*

$$\int_{\mathbf{R}^d} |T_1(f)(x)|^q v \, dx \le C_q \int_{\mathbf{R}^d} |T_2(f)(x)|^q M(v) \, dx,$$

where the constant C_q only depends on q. Then, for all $1 < p < \infty$ and all $w \in A_p$, there is a constant $C = C(w,p,d)$ such that

$$\int_{\mathbf{R}^d} |T_1(f)(x)|^p w \, dx \le C \int_{\mathbf{R}^d} |T_2(f)(x)|^p w \, dx$$

for all $f \in X$.

After we show that $L^p(w) \subset L^1_{\mathrm{loc}}(\mathbf{R}^d)$ for all $w \in A_p$ (see exercise 7.2), Theorem 7.2 will follow from Theorem 7.3 by setting $X = L^p(w)$, $T_1(f) = G_\beta(f)$, and $T_2(f) = f$.

We will now explore some properties of A_p weights. Much of this will involve review of material from earlier chapters. We hope the reader has been doing the exercises.

The next definition is redundant with (but also consistent with) a definition we gave earlier.

Definition 7.1. *A weight $v \in L^1_{\mathrm{loc}}(\mathbf{R}^d)$ belongs to the class A_∞ if, for all $\epsilon > 0$, there is a $\delta > 0$, such that, for all cubes Q and measurable $E \subset Q$,*

$$|E|/|Q| < \delta \Rightarrow v(E) \le \epsilon v(Q).$$

Lemma 7.1. *If $v \in A_\infty$, it is doubling.*

Proof of Lemma 7.1. Pick $\epsilon = 1/2$, find the appropriate $\delta > 0$, and let $\rho > 1$ be so close to 1 that $|\rho Q \setminus Q| < \delta |\rho Q|$ for all cubes Q; where ρQ means the concentric ρ-dilate of Q. Algebra then implies that $v(\rho Q) \le 2v(Q)$. Now let n be such that $\rho^n \ge 2$. Repeating this argument n times implies $v(2Q) \le 2^n v(Q)$.

We leave the proof of the next result to the reader:

Corollary 7.1. *Let v be a weight. If $v \in A_\infty$ and $v \not\equiv 0$, then $v(Q) > 0$ for all cubes Q.*

Theorem 7.4. *Let v be a weight, and suppose that $v(Q) > 0$ for all cubes Q. The following are equivalent: a) $v \in A_\infty$; b) there exist numbers η and τ, $0 < \eta, \tau < 1$, such that, for all cubes Q and measurable $E \subset Q$,*

$$|E|/|Q| \le \eta \Rightarrow v(E) \le \tau v(Q);$$

c) there exist numbers η and τ, $0 < \eta, \tau < 1$, such that, for all cubes Q and measurable $E \subset Q$,

$$v(E) \le \tau v(Q) \Rightarrow |E|/|Q| \le \eta.$$

Proof of Theorem 7.4. It is clear that a)$\Rightarrow$b). It is also not hard to see that b) and c) are equivalent. For example, suppose that b) holds for some τ and η. Let Q be a cube and let $E \subset Q$ satisfy $v(E) < (1 - \tau)v(Q)$. Such an E must exist, because $v(Q) > 0$. Then $v(Q \setminus E) > \tau v(Q)$, implying (by b)) $|Q \setminus E|/|Q| > \eta$, and therefore, $|E|/|Q| < 1 - \eta$, which implies c). A similar argument shows that c)$\Rightarrow$b). We conclude by showing b)$\Rightarrow$a). Fix the appropriate η and τ, and let $Q \equiv Q_0$ be a fixed cube, which we shall assume (without loss of generality) to be *dyadic*. We shall also assume that $v_{Q_0} = 1$. For $k = 1, 2, \ldots$, let $\{Q_j^k\}_j$ be the maximal dyadic cubes of Q_0 such that

$$v_{Q_j^k} > A^k,$$

where $A > 1$ is chosen so large that

$$\sum_j |Q_j^1| \leq \eta |Q_0|$$

and, for all $k \geq 1$ and j,

$$\sum_{j':Q_{j'}^{k+1} \subset Q_j^k} |Q_{j'}^{k+1}| \leq \eta |Q_j^k|,$$

both of which will hold if

$$\frac{2^d}{A} \leq \eta.$$

These inequalities imply that

$$\sum_j v(Q_j^1) \leq \tau v(Q_0)$$

and, for all k and j,

$$\sum_{j':Q_{j'}^{k+1} \subset Q_j^k} v(Q_{j'}^{k+1}) \leq \tau v(Q_j^k).$$

Induction on k now implies that

$$\sum_j v(Q_j^k) \leq \tau^k v(Q_0)$$

holds for all k. Because of the Lebesgue Differentiation Theorem, we now have, for all $k \geq 1$,

$$v\left(\{x \in Q_0 : v(x) > A^k\}\right) \leq \tau^k v(Q_0),$$

with the consequence that, for any $\epsilon > 0$,

$$\frac{1}{|Q_0|} \int_{Q_0} (v(x))^{1+\epsilon}\, dx \leq A^{1+\epsilon} + \sum_1^\infty \int_{x \in Q_0:A^k < v(x) \leq A^{k+1}} (v(x))^{1+\epsilon}\, dx$$

$$\leq A^{1+\epsilon} + C\sum_1^\infty A^{k\epsilon} \int_{x \in Q_0:A^k < v(x) \leq A^{k+1}} v(x)\, dx$$

$$\leq A^{1+\epsilon} + C\sum_1^\infty A^{k\epsilon} \tau^k$$

$$\leq C,$$

if ϵ is chosen so small that $A^\epsilon \tau < 1$. We proved this under the assumption that $v_{Q_0} = 1$. By homogeneity, we can immediately infer the existence of positive numbers ϵ and K such that, for all cubes Q,

$$\left(\frac{1}{|Q|} \int_Q (v(x))^{1+\epsilon}\, dx\right)^{1/(1+\epsilon)} \leq \frac{K}{|Q|} \int_Q v(x)\, dx.$$

This is the reverse Hölder inequality we saw earlier. Let $E \subset Q$ be measurable. Using the "forward" and reverse Hölder inequalities together, we get:

$$\frac{1}{|Q|} \int_E v(x)\,dx = \frac{1}{|Q|} \int_Q v(x)\,\chi_E(x)\,dx$$

$$\leq \left(\frac{1}{|Q|} \int_Q (v(x))^{1+\epsilon}\,dx\right)^{1/(1+\epsilon)} (|E|/|Q|)^\epsilon$$

$$\leq \left(\frac{K}{|Q|} \int_Q v(x)\,dx\right) (|E|/|Q|)^\epsilon,$$

yielding

$$v(E) \leq K v(Q)(|E|/|Q|)^\epsilon,$$

which implies a).

The preceding theorem had to assume that $v(Q) > 0$ for all cubes Q. The next lemma shows that, unless $v \equiv 0$, such a hypothesis is unnecessary.

Lemma 7.2. *Let v be a weight and suppose that $v \not\equiv 0$. If v satisfies a), b), or c) from Theorem 7.4, then $v(Q) > 0$ for all cubes Q.*

Proof of Lemma 7.2. We already know this for a). Suppose that c) holds, and let Q be any cube. Starting with $Q \equiv Q_0$, build a tower of nested cubes,

$$Q_0 \subset Q_1 \subset Q_2 \subset \ldots Q_N,$$

such that $v(Q_N) > 0$ and $|Q_i|/|Q_{i+1}| > \eta$ for each $0 \leq i < N$. Working downwards from Q_N, we get $v(Q_0) > \tau^N v(Q_N) > 0$. For b), we use a similar tower of cubes, but this time choose them so that, for each $0 \leq i < N$,

$$|Q_{i+1} \setminus Q_i| \leq \eta |Q_{i+1}|,$$

which ensures that

$$v(Q_{i+1} \setminus Q_i) \leq \tau v(Q_{i+1}),$$

and thus

$$v(Q_i) \geq (1 - \tau)v(Q_{i+1}),$$

leading to

$$v(Q_0) \geq (1 - \tau)^N v(Q_N) > 0.$$

Henceforth we will assume that every weight we consider is non-trivial.

Lemma 7.2 and the proof of Theorem 7.4 yield this corollary, with no extra work.

Corollary 7.2. *A weight v belongs to A_∞ if and only if it satisfies the reverse Hölder inequality for some K and ϵ.*

Corollary 7.3. *For all* $1 < p < \infty$, $A_p \subset A_\infty$.

Proof of Corollary 7.3. Set $R = \|v\|_{A_p}$, and fix a cube Q. For $A > 1$, define
$$E_A = \{x \in Q : \; v(x) \le A^{-1} v_Q\}.$$
An easy computation shows that
$$\left(\frac{1}{|Q|} \int_Q v(x)\, dx\right)^{1/p} \left(\frac{1}{|Q|} \int_{E_A} v(x)^{1-p'}\, dx\right)^{1/p'} \ge A^{1/p}(|E_A|/|Q|)^{1/p'}.$$

However, the left-hand side of the preceding inequality is less than or equal to R. Therefore,
$$|E_A|/|Q| \le A^{1-p'} R^{p'}.$$
Let A be so large that $A^{1-p'} R^{p'} \le 1/4$. If $E \subset Q$ and $|E|/|Q| > 1/2$, then $v(x) > A^{-1} v_Q$ on at least half of E, which means that $v(x) > A^{-1} v_Q$ on a subset with measure $\ge (1/4)|Q|$. Therefore $v(E) > (4A)^{-1} v(Q)$, which is c) from Theorem 7.4. Therefore $v \in A_\infty$.

When we combine the preceding corollary with Theorem 5.5 (note the *last* sentence of its statement), we obtain:

Corollary 7.4. *Let* $1 < p < \infty$ *and* $w \in A_p$. *Suppose that* $0 < \beta \le 1$. *There is a constant* $C = C(w, p, d, \beta)$ *such that, for all* $f \in \cup_{1 \le r < \infty} L^r(dx)$ *(the usual "unweighted" L^r spaces),*
$$\int_{\mathbf{R}^d} |f(x)|^p\, w\, dx \le C \int_{\mathbf{R}^d} (G_\beta(f)(x))^p\, w\, dx.$$

Proof of Corollary 7.4. Since $w \in A_\infty$, Theorem 5.5 implies
$$\int_{\mathbf{R}^d} |f(x)|^p\, w\, dx \le C \int_{\mathbf{R}^d} (\tilde{S}_{sd}(f)(x))^p\, w\, dx,$$
where $\tilde{S}_{sd}(f)$ is as we defined it in 5.9. But $\tilde{S}_{sd}(f)(x) \le C G_\beta(f)(x)$ pointwise.

Remark. The reader should be aware of what Corollary 7.4 does not say. It does *not* say that
$$\int_{\mathbf{R}^d} |f(x)|^p\, w\, dx \le C \int_{\mathbf{R}^d} (G_\beta(f)(x))^p\, w\, dx \tag{7.5}$$
for $f \in L^p(w)$. For example, it is possible for f to belong to $L^2(w)$, with w an A_2 weight, but not belong to $L^r(dx)$ for any $1 \le r \le \infty$. (See exercise 7.5.) Nevertheless, inequality 7.5 does hold for $f \in L^p(w)$ when $w \in A_p$ $(1 < p < \infty)$; see Corollary 7.7 at the end of this chapter.

Strictly speaking, we won't need the next result for our Littlewood-Paley results. We are including it here because it is a standard part of the theory of A_p weights.

Corollary 7.5. *If $v \in A_\infty$, then $v \in A_p$ for some $1 < p < \infty$.*

Proof of Corollary 7.5. Let $Q = Q_0$ be a cube, which we will take to be dyadic. We assume that $v_{Q_0} = 1$. Let $A > 1$ be a large number (to be specified presently). For $k = 1, 2, \ldots$, let $\{Q_j^k\}_j$ be the maximal dyadic subcubes of Q_0 such that

$$v_{Q_j^k} < A^{-k}.$$

Because v is doubling, these cubes will also satisfy

$$v_{Q_j^k} \geq cA^{-k}$$

for some fixed $c > 0$. Therefore, by taking A big enough, we can ensure that, for every Q_j^k,

$$\sum_{j': Q_{j'}^{k+1} \subset Q_j^k} v(Q_{j'}^{k+1}) \leq \tau v(Q_j^k),$$

and also that

$$\sum_j v(Q_j^1) \leq \tau v(Q_0),$$

where τ is the number from statement c) in Theorem 7.4. Therefore, for each Q_j^k,

$$\sum_{j': Q_{j'}^{k+1} \subset Q_j^k} |Q_{j'}^{k+1}| \leq \eta |Q_j^k|$$

and

$$\sum_j |Q_j^1| \leq \eta |Q_0|;$$

which, by induction, imply

$$\sum_j |Q_j^k| \leq \eta^k |Q_0|$$

for all $k \geq 1$. By Lebesgue's theorem again, the subset of Q where $v(x) < A^{-k}$ is almost-everywhere contained in $\cup_j Q_j^k$. Therefore, for any $\epsilon > 0$,

$$\frac{1}{|Q|} \int_Q v(x)^{-\epsilon} \, dx \leq A^\epsilon + \left(\sum_1^\infty A^{k\epsilon} \eta^k \right)$$

$$\leq C,$$

if we choose ϵ so small that $A^\epsilon \eta < 1$. That proves that $v \in A_p$ if $p' - 1 < \epsilon$.

Corollary 7.6. *If $v \in A_\infty$ then $\{x : v(x) = 0\}$ has Lebesgue measure 0.*

The next theorem provides the reason the A_p classes can be considered "natural." It is also the key to proving Theorem 7.3. We will state the theorem, discuss it briefly, state and prove a lemma, and then prove the theorem.

Theorem 7.5. *Let* $1 < p < \infty$. *If* $w \in A_p$, *the Hardy-Littlewood maximal operator* $M(\cdot)$ *is bounded on* $L^p(w)$; *i.e., there is a constant* $C = C(p, w)$ *such that, for all* $f \in L^p(w)$,

$$\int_{\mathbf{R}^d} (M(f)(x))^p \, w \, dx \leq C \int_{\mathbf{R}^d} |f(x)|^p \, w \, dx. \tag{7.6}$$

Remark. This theorem is due to Benjamin Muckenhoupt [42]. The A_p condition is also necessary for 7.6. To (almost) see this, put $f = w^{\alpha} \chi_Q$ and plug it into 7.6. Out pops:

$$\int_Q \left(\frac{1}{|Q|} \int_Q w^{\alpha} \, dt \right)^p w \, dx \leq C \int_Q w^{\alpha p + 1} \, dx.$$

Setting $\alpha = 1 - p'$ (which has the effect of making $\alpha = \alpha p + 1$) and canceling yields:

$$\left(\frac{1}{|Q|} \int_Q w^{1-p'} \, dt \right)^{p-1} \left(\frac{1}{|Q|} \int_Q w \, dx \right) \leq C,$$

which yields the A_p condition, *provided we know that* $w^{\alpha} \chi_Q \in L^p(w)$. How to fill this small gap is left as an exercise.

Lemma 7.3. *Let* $1 < p < \infty$. *If* $w \in A_p$ *then there exists an* r, $1 < r < p$, *such that* $w \in A_r$.

Proof of Lemma 7.3. If $w \in A_p$ then $w^{1-p'} \in A_{p'} \subset A_{\infty}$, implying that, for some positive ϵ and K,

$$\left(\frac{1}{|Q|} \int_Q w(x)^{(1-p')(1+\epsilon)} \, dx \right)^{1/(1+\epsilon)} \leq \frac{K}{|Q|} \int_Q w(x)^{1-p'} \, dx$$

holds for all cubes Q. Define r by the equation $(1 - p')(1 + \epsilon) = 1 - r'$; which, after some algebra, yields

$$1 < r = \frac{p + \epsilon}{1 + \epsilon} < p.$$

I claim that, for any cube Q,

$$\left(\frac{1}{|Q|} \int_Q w(x) \, dx \right)^{1/r} \left(\frac{1}{|Q|} \int_Q w(x)^{1-r'} \, dx \right)^{1/r'} \leq C, \tag{7.7}$$

where C is a constant not depending on Q. Proving 7.7 is really just a matter of unraveling the exponents. The left-hand side of 7.7 is

$$\left(\frac{1}{|Q|} \int_Q w(x) \, dx \right)^{(1+\epsilon)/(p+\epsilon)} \left(\frac{1}{|Q|} \int_Q w(x)^{(1-p')(1+\epsilon)} \, dx \right)^{(p-1)/(p+\epsilon)}.$$

Because $w^{1-p'}$ satisfies the reverse Hölder inequality for $1 + \epsilon$ and K, the product is bounded by a constant times

$$\left(\frac{1}{|Q|} \int_Q w(x) \, dx \right)^{(1+\epsilon)/(p+\epsilon)} \left(\frac{1}{|Q|} \int_Q w(x)^{(1-p')} \, dx \right)^{(1+\epsilon)(p-1)/(p+\epsilon)} .$$

When we take a $(1 + \epsilon)/(p + \epsilon)$ root of this, we get

$$\left(\frac{1}{|Q|} \int_Q w(x) \, dx \right) \left(\frac{1}{|Q|} \int_Q w(x)^{(1-p')} \, dx \right)^{p-1} ;$$

and if we take a pth root of *that*, we get

$$\left(\frac{1}{|Q|} \int_Q w(x) \, dx \right)^{1/p} \left(\frac{1}{|Q|} \int_Q w(x)^{1-p'} \, dx \right)^{1/p'} \leq \|w\|_{A_p} .$$

The lemma is proved.

Proof of Theorem 7.5. We will prove $L^p(w)$ boundedness for $M_d(\cdot)$. The doubling property of w will then imply the existence of two positive constants c_1 and c_2 such that, for all $\lambda > 0$,

$$w \left(\{x : \ M(f)(x) > \lambda\} \right) \leq c_1 w \left(\{x : \ M_d(f)(x) > c_2 \lambda\} \right).$$

We leave this implication as an exercise; it follows the same pattern as with Lebesgue measure (see exercise 2.3.) For $T > 0$, define

$$M_{d,T}(f)(x) \equiv \sup_{\substack{x \in Q \in \mathcal{D}_d \\ \ell(Q) \leq T}} \frac{1}{|Q|} \int_Q |f(t)| \, dt.$$

By Monotone Convergence, we only need to show that $M_{d,T}(\cdot)$ maps $L^p(w) \mapsto L^p(w)$, with a bound independent of T. Since $M_{d,T}(\cdot)$ is trivially bounded on $L^\infty(w)$, L^p boundedness will follow if we can show that $M_{d,T}(\cdot)$ is weak-type (r, r) for some $1 < r < p$ (with weak-type bounds independent of T, of course). However, we know that there exists an r $(1 < r < p)$, such that $w \in A_r$. Therefore our theorem will follow from this: *If $1 < r < \infty$ and $w \in A_r$, there exists a constant $C = C(w, d, r)$ such that, for all $f \in L^r(w)$ and $\lambda > 0$,*

$$w \left(\{x : \ M_{d,T}(f)(x) > \lambda\} \right) \leq \frac{C}{\lambda^r} \int |f(x)|^r \, w \, dx. \tag{7.8}$$

We finish by showing 7.8.

Because it will be convenient, let us set $\sigma(x) = w(x)^{1-r'}$. If Q is any cube, then

$$\frac{1}{|Q|} \int_Q |f(x)| \, dx = \frac{1}{|Q|} \int_Q |f(x)| \, w(x)^{1/r} \, w(x)^{-1/r} \, dx$$

$$\leq \left(\frac{1}{|Q|} \int_Q |f(x)|^r \, w \, dx \right)^{1/r} \left(\frac{1}{|Q|} \int_Q \sigma(x) \, dx \right)^{1/r'} .$$

But

$$\left(\frac{1}{|Q|} \int_Q \sigma(x)\,dx \right)^{1/r'} \le C \left(\frac{1}{|Q|} \int_Q w(x)\,dx \right)^{-1/r},$$

because $w \in A_r$. Therefore:

$$\frac{1}{|Q|} \int_Q |f(x)|\,dx \le C \left(\frac{1}{w(Q)} \int_Q |f(x)|^r w\,dx \right)^{1/r}. \tag{7.9}$$

Let $\{Q_k^\lambda\}_k$ be the maximal dyadic cubes such that $\ell(Q_k^\lambda) \le T$ and

$$\frac{1}{|Q_k^\lambda|} \int_Q |f(x)|\,dx > \lambda.$$

(The restriction to cubes satisfying $\ell(Q) \le T$ is an easy way to ensure that these *maximal* cubes exist.) The union of the Q_k^λ's is the set where $M_{d,T}(f)$ is bigger than λ. Because of 7.9, each of these cubes satisfies:

$$w(Q_k^\lambda) \le \frac{C}{\lambda^r} \int_{Q_\lambda^k} |f(x)|^r w\,dx.$$

Summing the preceding inequality over k yields 7.8. We have proved Theorem 7.5.

Proof of Theorem 7.3. Let $1 < q < \min(p, p_0)$, where q is to be specified presently. Let g be a non-negative function satisfying

$$\int (g(x))^{(p/q)'} w\,dx = 1.$$

What we need to estimate is

$$\int_{\mathbf{R}^d} |T_1(f)(x)|^q g(x) w\,dx.$$

By hypothesis, this is bounded by a constant times

$$\int_{\mathbf{R}^d} |T_2(f)(x)|^q M(v)\,dx, \tag{7.10}$$

where $v = g(x)w$. We re-write 7.10 as

$$\int_{\mathbf{R}^d} \left(|T_2(f)(x)|^q w^{q/p} \right) \left(M(v)\,w^{-q/p} \right)\,dx.$$

By Hölder's inequality, this is less than or equal to the product of

$$\left(\int_{\mathbf{R}^d} |T_2(f)(x)|^p w\,dx \right)^{q/p}$$

and

$$\left(\int_{\mathbf{R}^d} (M(v)(x))^{p/(p-q)} \, w^{-q/(p-q)} \, dx \right)^{(p-q)/p} . \qquad (7.11)$$

The first factor is just fine: it equals $\|T_2(f)\|_{L^p(w)}^q$. We will conclude the proof by showing that, for appropriate q (depending on p and w), the second factor is bounded by the "constantly changing constant."

Pick q so close to 1 that $w \in A_{p/q}$. Then $w^{1-(p/q)'} \in A_{(p/q)'}$. In plain language, this means that

$$w^{-q/(p-q)} \in A_{p/(p-q)}.$$

But now Theorem 7.5 implies that 7.11 is bounded by a constant times

$$\left(\int_{\mathbf{R}^d} (v(x))^{p/(p-q)} \, w^{-q/(p-q)} \, dx \right)^{(p-q)/p} .$$

After unpacking the definition of v, that last expression turns out to be:

$$\left(\int_{\mathbf{R}^d} (g(x))^{(p/q)'} \, w(x)^{p/(p-q)} \, w(x)^{-q/(p-q)} \, dx \right)^{(p-q)/p} ,$$

which simplifies to

$$\left(\int_{\mathbf{R}^d} (g(x))^{(p/q)'} \, w(x) \, dx \right)^{(p-q)/p} = 1.$$

Except for the few loose ends we have left as exercises, Theorem 7.3 is proved.

Much as in the unweighted case, the preceding arguments give us the following corollary, whose proof we leave to the reader.

Corollary 7.7. *Let $1 < p < \infty$ and $w \in A_p$. Suppose that $0 < \beta \leq 1$. There is a constant $C = C(p, w, d, \beta)$ such that, for all $f \in L^p(w)$,*

$$\int_{\mathbf{R}^d} |f(x)|^p \, w \, dx \leq C \int_{\mathbf{R}^d} (G_\beta(f)(x))^p \, w \, dx.$$

Exercises

7.1. Show that, for all $1 < p < q < \infty$, $A_p \subset A_q$.

7.2. a) Show that, if $w \in A_p$ $(1 < p < \infty)$, then $(1 + |x|)^{-d} \in L^p(w)$. b) Show that, if $w \in A_p$ $(1 < p < \infty)$ and $f \in L^p(w)$, then $|f|(1 + |x|)^{-d} \in L^1(dx)$ (the usual "unweighted" L^1). Deduce that $L^p(w) \subset L^1_{\mathrm{loc}}(\mathbf{R}^d)$. c) Show that, if $w \in A_p$ $(1 < p < \infty)$, $f \in L^p(w)$, $\phi \in \mathcal{C}_{(\beta,\epsilon)}$, and K is a measurable set with compact closure contained in $\mathbf{R}_+^{d+1}$, then $f * \phi_y(t)$ is defined and *bounded* on K. Hint for b): Begin by writing

$$|f|(1 + |x|)^{-d} = (|f|w^{1/p})((1 + |x|)^{-d}w^{-1/p}).$$

7.3. As $p \searrow 1$, the A_p condition naturally morphs into

$$\sup_{Q \subset \mathbf{R}^d} \left(\frac{1}{|Q|} \int_Q w(x)\, dx \right) \|w^{-1}\|_{L^\infty(Q)} < \infty. \tag{7.12}$$

A weight is said to belong to A_1 if 7.12 holds; the supremum, denoted by $\|w\|_{A_1}$, is called w's A_1 norm. a) Show that $w \in A_1$ if and only if there is a constant C such that $M(w)(x) \le Cw(x)$ almost everywhere. b) Show that $A_1 \subset A_p$ for all $p > 1$, and that $w \in A_1$ implies that the Hardy-Littlewood maximal operator is weak-type $(1,1)$ with respect to w; i.e., that there is a constant C such that, for all $f \in L^1(w)$ and all $\lambda > 0$,

$$w\left(\{x : M(f)(x) > \lambda\}\right) \le \frac{C}{\lambda} \int_{\mathbf{R}^d} |f(x)|\, w\, dx. \tag{7.13}$$

(See exercise 3.5 and the results it refers to.) c) Show that, if $M(\cdot)$ is weak-type $(1,1)$ with respect to w (i.e., if 7.13 holds), then $w \in A_1$. (Hint: Let $x_0 \in \mathbf{R}^d$ be such that

$$w(x_0) = \lim_{\substack{\ell(Q) \to 0 \\ x_0 \in Q}} \frac{1}{|Q|} w(Q),$$

and test the weak-type inequality against the functions of the form $|Q|^{-1} \chi_Q(x)$, where the cubes Q are homing in on x_0.)

7.4. Let $q : \mathbf{R}^d \mapsto \mathbf{R}$ be a polynomial, and suppose that $q \not\equiv 0$. In exercise 3.1 you were asked to show that $|q| \in A_\infty$. Now show that, for all $1 < p < \infty$, there is a $\delta > 0$ such that $|q|^\delta$ and $|q|^{-\delta}$ both belong to A_p. Show that, unless q is constant, no strictly positive power of $|q|$ can belong to A_1.

7.5. Define $g : \mathbf{R} \mapsto \mathbf{R}$ by

$$g(x) \equiv |x|^{1/2} \chi_{[-1/2, 1/2]}(x),$$

and set

$$w(x) = \sum_{-\infty}^{\infty} g(x - n).$$

Show that $w \in A_2$. For $n = 1, 2, 3, \ldots$, define $c_n = n^{-1/2}$, $\epsilon_n = 1/\log(10+n)$, and let I_n be the open interval $(n - \epsilon_n, n + \epsilon_n)$. Define

$$f(x) \equiv |x|^{-1/2} \chi_{[-1/2, 1/2]}(x) + \left(\sum_1^\infty c_n \chi_{I_n}(x) \right).$$

Show that $f \in L^2(w)$, but that f does not belong to $L^r(\mathbf{R}, dx)$ for any $1 \le r \le \infty$.

7.6. Let $w \in A_p$ $(1 < p < \infty)$ and suppose that $f \in L^p(w)$. Show directly that the $L^p(w)$ limit of $\{f_{(K_i),\phi,\psi}\}$ (where $\{K_i\}$ is a compact-measurable exhaustion) is independent of the exhaustion. (Hint: Let $\{L_i\}$ be any other compact-measurable exhaustion. Show that $g_{\phi,K_i\Delta L_i}(f) \to 0$ in $L^p(w)$, and consider what this says about $\|f_{(K_i),\phi,\psi} - f_{(L_i),\phi,\psi}\|_{L^p(w)}$.)

7.7. Following the lead given in the text, show that the A_p condition is necessary for the $L^p(w)$ boundedness of the Hardy-Littlewood maximal operator. (Hint: Consider weights of the form $w(x)+\epsilon$ as $\epsilon \searrow 0$.) A very similar argument will now show that the A_p condition is necessary for the $L^p(w)$ boundedness of the intrinsic square function. (Hint: Consider $f = w^\alpha \chi_Q$, and get a lower bound on $\tilde{A}_{(\beta,\epsilon)}(f)(t,y)$ when (t,y) doesn't stray too far from $T(Q)$.)

7.8. Show that, if $w \in L^1_{\text{loc}}(\mathbf{R}^d)$ is any doubling weight, then there exist positive constants c_1 and c_2 such that, for all $\lambda > 0$ and all $f \in L^1_{\text{loc}}(\mathbf{R}^d)$,

$$w\left(\{x: \ M(f)(x) > \lambda\}\right) \leq c_1 w\left(\{x: \ M_d(f)(x) > c_2\lambda\}\right).$$

7.9. To what extent—if any—do the analogous equivalences of Theorem 7.4 hold in the *dyadic* setting?

7.10. a) Show that, if $w \in L^1_{\text{loc}}(\mathbf{R}^d)$ is any weight and $0 < p < \infty$, then $\mathcal{C}_0^\infty(\mathbf{R}^d)$ is dense in $L^p(w)$ (no Littlewood-Paley theory needed). b) Let $\phi \in \mathcal{C}_0^\infty(\mathbf{R}^d)$ satisfy $\int \phi \, dx = 1$. For $f \in L^1_{\text{loc}}(\mathbf{R}^d)$ and $y > 0$, define

$$f(x,y) = f * \phi_y(x).$$

Show that, if $w \in A_p$ $(1 < p < \infty)$ and $f \in L^p(w)$, then $f(\cdot,y) \to f$ in $L^p(w)$ as $y \to 0$. The upshot is that, if $w \in A_p$, then smooth, L^p approximations to $f \in L^p(w)$ can be obtained in this canonical fashion.

7.11. Prove this generalization of Theorem 7.3. Suppose that X is a non-empty set, and T_1 and T_2 are two operators mapping from X into $L^1_{\text{loc}}(\mathbf{R}^d)$. Suppose there exists a number $p_0 > 1$ such that, for all $1 < q < p_0$, we can find a $k = k(q)$ with the following property: For all weights v, and all $f \in X$,

$$\int_{\mathbf{R}^d} |T_1(f)(x)|^q \, v \, dx \leq C_q \int_{\mathbf{R}^d} |T_2(f)(x)|^q \, M^k(v) \, dx,$$

where the constant C_q only depends on q, and $M^k(\cdot)$ means a k-fold iteration of the maximal operator $M(\cdot)$. Show: For every $1 < p < \infty$ and every $w \in A_p$, there is a constant $C = C(w,p,d)$ such that, for all $f \in X$,

$$\int_{\mathbf{R}^d} |T_1(f)(x)|^p \, w \, dx \leq C \int_{\mathbf{R}^d} |T_2(f)(x)|^p \, w \, dx.$$

Notes

Our analysis of A_∞ is closely based on the discussion in [13]. The extrapolation trick (Theorem 7.3) is implicit in arguments in [16] and [24]. In fact, a stronger result is true: If, for all $1 < q < p_0$ and all A_1 weights w, there is a constant $C(q, w)$ such that

$$\int_{\mathbf{R}^d} |T_1(f)(x)|^p \, w \, dx \leq C \int_{\mathbf{R}^d} |T_2(f)(x)|^p \, w \, dx$$

for all $f \in X$, then, for all $1 < p < \infty$ and all $w \in A_p$, there is a constant $C(p, w)$ such that

$$\int_{\mathbf{R}^d} |T_1(f)(x)|^p \, w \, dx \leq C(p, w) \int_{\mathbf{R}^d} |T_2(f)(x)|^p \, w \, dx$$

holds for all $f \in X$. (These facts were pointed out to the author by D. Cruz-Uribe and J. Martell at a conference in Sevilla.) The second fact is the key to the celebrated *Rubio de Francia Extrapolation Theorem*: If, for some *fixed* $1 < p_0 < \infty$, a sublinear operator T maps boundedly from $L^{p_0}(w)$ into itself, for all $w \in A_{p_0}$, then T maps boundedly from $L^p(w)$ into itself, for *all* $w \in A_p$, for *all* $1 < p < \infty$ ([49]; see [16], pp. 141–142, for a nice proof). We believe that C. Pérez was the first to observe that the extrapolation trick could be extended to arbitrary (and not necessarily sublinear) *pairs* of operators T_1 and T_2. The analogue of Theorem 7.2, for the classical Lusin square function on the unit disk, was first proved by C. Segovia and R. L. Wheeden in [50]. The first proof for the the Lusin square function in $\mathbf{R}^{d+1}_+$ is due to R. F. Gundy and R. L. Wheeden [26]. These classical proofs used good-λ inequalities. They were generalized to a particular "real-variable only" square function in [35]. Extensions to more general real-variable square functions appear in [55] and [56]. Theorem 7.1 is something that "everybody knows," but which, to our knowledge, has not previously been proved.

8

Schrödinger Operators

This chapter will have a long lead-up to a short payoff.

The reader might want to refresh his memory concerning Theorem 4.6. We will be using it soon.

In non-relativistic quantum mechanics, a particle is represented by a *wave function* $u : \mathbf{R}^3 \times (-\infty, \infty) \mapsto \mathbf{C}$, such that $|u(x,t)|^2$ is the probability density of finding the particle in the neighborhood of x at time t. More precisely, if $E \subset \mathbf{R}^3$ is measurable, then

$$\int_E |u(x,t)|^2 \, dx$$

is the probability that the particle will be in the set E at time t.

The wave function evolves in time according to the Schrödinger equation, which we we may write as:

$$i\frac{\partial u}{\partial t} = \mathcal{H}u, \tag{8.1}$$

where $\mathcal{H}$ is the Hamiltonian operator. If the particle is moving in a potential field $-V(x)$ that only depends on x, then, with appropriate choices of units, the operator $\mathcal{H}$ can be written $\mathcal{H} = -\Delta - V(x)$, where this means:

$$\mathcal{H}u(x,t) \equiv -\Delta_x u(\cdot, t) - V(x)u(x,t).$$

The Hamiltonian's two parts, $-\Delta$ and $-V$, represent, respectively, the particle's kinetic and potential energies. In our discussion, we will only consider V's that are non-negative. This has the effect of turning the Hamiltonian's $-V$ into a potential well—i.e., one that can trap a particle.

The eigenfunctions of $\mathcal{H}$ are identified with the particle's possible energy states, with the eigenvalues being the energies. According to quantum mechanical formalism, the energy of the wave function u is given by the inner product $\langle \mathcal{H}u, u \rangle$, or

$$\int_{\mathbf{R}^3} (-\Delta_x u(x,t) - V(x)u(x,t))\bar{u}(x,t) \, dx.$$

Integrating by parts in x, this seen to be formally equal to

$$\int_{\mathbf{R}^3} \left(|\nabla_x u(x,t)|^2 - V(x)|u(x,t)|^2 \right) dx.$$

We are most interested in whether this quantity can ever be negative for some u, which would physically correspond to the potential $-V$ trapping a particle.

We will henceforth ignore any possible dependence u might have on t, and only consider functions u depending on $x \in \mathbf{R}^3$. We shall also only consider functions u which are infinitely differentiable and have compact support.

With these assumptions, we can now phrase the main mathematical question of this chapter: What conditions on a non-negative, locally integrable V imply that

$$\int_{\mathbf{R}^3} |u(x)|^2 \, V(x) \, dx \leq \int_{\mathbf{R}^3} |\nabla u(x)|^2 \, dx \tag{8.2}$$

holds for all $u \in \mathcal{C}_0^\infty(\mathbf{R}^3)$? Given our previous discussion, if 8.2 always holds, then $-V$ can never trap a particle.

We can easily find a necessary condition for 8.2. Let ϕ be a fixed function in $\mathcal{C}_0^\infty(\mathbf{R}^3)$ satisfying:

$$\phi(x) = \begin{cases} 1 & \text{if } |x| \leq 1; \\ 0 & \text{if } |x| > 1.1. \end{cases}$$

Set $R = \int_{\mathbf{R}^3} |\nabla \phi|^2 \, dx$. Applying 8.2 to ϕ, we see that a necessary condition is

$$V\left(\{x : |x| \leq 1\}\right) \leq R,$$

which is not very interesting. Now let $\delta > 0$ and apply 8.2 to the function $\phi(\cdot/\delta)$. The left-hand side of 8.2 becomes the V-measure of $\{x : |x| < \delta\}$, while the left-hand side is now δR. Thus, a necessary condition for 8.2 is that

$$V\left(\{x : |x| < \delta\}\right) \leq \delta R$$

for all $\delta > 0$. By considering translations of ϕ, we get our "trivial" necessary condition for 8.2: There is a constant R such that, in order for 8.2 to hold, we must have

$$V(B(t; \delta)) \leq R\delta \tag{8.3}$$

for all balls $B(t; \delta) \subset \mathbf{R}^3$.

The reader will not have a hard time showing that 8.3 implies the following: There is a positive constant C such that, in order for 8.2 to hold, we must have

$$\frac{1}{|Q|} \int_Q V(x) \, dx \leq C\ell(Q)^{-2} \tag{8.4}$$

all cubes $Q \subset \mathbf{R}^3$.

Inequality 8.4 is in some ways more natural than 8.3. If V is fairly smooth and Q is not too big, then $V(x)$ will stay close to its average for $x \in Q$, and then 8.4 is saying that V does not have any deep wells. Put another way, it says that, in order to trap a particle in a cube Q, you need a potential well of depth roughly $\ell(Q)^{-2}$. This is a semi-quantitative statement of the Heisenberg uncertainty principle. If we confine a particle in a region of diameter $\sim\ell(Q)$, the uncertainty in its momentum must be $\sim\ell(Q)^{-1}$, forcing the expected value of its kinetic energy to be $\sim\ell(Q)^{-2}$. If the potential well isn't deep enough, the particle's kinetic energy will allow it to escape.

Inequality 8.4 is not sufficient for 8.2; a proof of this is sketched in the exercises. But, using weighted Littlewood-Paley theory, we can obtain a sufficient condition for 8.2 that is just a little stronger than 8.4.

Let us recall that we have defined the Fourier transform this way:

$$\hat{f}(\xi) = \int_{\mathbf{R}^3} f(x)\, e^{-2\pi i x \cdot \xi}\, dx,$$

where here we are assuming that $f \in \mathcal{C}_0^\infty(\mathbf{R}^3)$. Our convention has the consequence that

$$\int_{\mathbf{R}^3} |\nabla f(x)|^2\, dx = 4\pi^2 \int_{\mathbf{R}^3} |\xi|^2 |\hat{f}(\xi)|^2\, d\xi,$$

for the same f's.

Now let ϕ be a fixed real, radial function in the Schwartz class on $\mathbf{R}^3$, such that $\hat{\phi}(\xi) \geq 0$ for all ξ, and that also satisfies

$$\hat{\phi}(\xi) = \begin{cases} 1 & \text{if } 3/4 < |\xi| < 4/3; \\ 0 & \text{if } |\xi| \leq 1/2 \text{ or } |\xi| > 2. \end{cases}$$

Now let $\psi \in \mathcal{C}_0^\infty(\mathbf{R}^3)$ be real, radial, supported in $\{x : |x| \leq 1\}$, have integral equal to 0, and be normalized so that

$$\int_0^\infty \hat{\phi}(y\xi)\, \hat{\psi}(y\xi)\, \frac{dy}{y} \equiv 1$$

for all $\xi \neq 0$. (We leave it as an exercise to show that, given a ϕ as we have described, such a ψ always exists.)

If $f \in \mathcal{C}_0^\infty(\mathbf{R}^3)$ then our work on the Calderón reproducing formula lets us write

$$f(x) = \int_{\mathbf{R}_+^4} (f * \phi_y(t))\, \psi_y(x - t)\, \frac{dt\, dy}{y},$$

as the L^p limit (for all $1 < p < \infty$) of the sequence of functions

$$\int_{K_i} (f * \phi_y(t))\, \psi_y(x - t)\, \frac{dt\, dy}{y},$$

for any compact-measurable exhaustion $\{K_i\}_i$ of $\mathbf{R}_+^4$.

Our previous work (in particular, Theorem 4.6) implies that, for all $\tau > 1$, there is a constant $C = C(\tau)$ such that, for all $f \in \mathcal{C}_0^\infty(\mathbf{R}^3)$ and all non-negative, locally integrable V,

$$\int_{\mathbf{R}^3} |f(x)|^2\, V(x)\, dx \le C \sum_{Q \in \mathcal{D}_3} \frac{|\lambda_Q(f)|^2}{|\tilde{Q}|} \int_{\tilde{Q}} V(x)\, (\log(e + V(x)/V_{\tilde{Q}}))^\tau\, dx,$$

(8.5)

where

$$\lambda_Q(f) \equiv \left(\int_{T(Q)} |f * \phi_y(t)|^2 \frac{dt\, dy}{y} \right)^{1/2}.$$

If V satisfies

$$\frac{1}{|Q|} \int_Q V(x)\, (\log(e + V(x)/V_Q))^\tau\, dx \le \ell(Q)^{-2}$$

for all cubes $Q \subset \mathbf{R}^3$, then the right-hand side of 8.5 is less than or equal to a constant times

$$\sum_{Q \in \mathcal{D}_3} \ell(Q)^{-2} \int_{T(Q)} |f * \phi_y(t)|^2 \frac{dt\, dy}{y} \le C \int_{\mathbf{R}^4_+} y^{-2} |f * \phi_y(t)|^2 \frac{dt\, dy}{y}.$$

(8.6)

By Plancherel's Theorem (and Fourier convolution), the right-hand side of 8.6 is equal to

$$C \int_{\mathbf{R}^4_+} |\hat{f}(\xi)|^2 \left(\int_0^\infty |\hat{\phi}(y\xi)|^2 \frac{dy}{y^3} \right) d\xi.$$

A change of variable and our hypothesis on the support of $\hat{\phi}$ imply that

$$\int_0^\infty |\hat{\phi}(y\xi)|^2 \frac{dy}{y^3} \equiv c_\phi |\xi|^2$$

for all ξ, for some constant $c_\phi > 0$. Therefore the right-hand side of 8.6 is less than or equal to a constant times

$$\int_{\mathbf{R}^3} |\xi|^2 |\hat{f}(\xi)|^2\, d\xi \le C \int_{\mathbf{R}^3} |\nabla f(x)|^2\, dx.$$

Putting everything together, we have the following theorem:

Theorem 8.1. *Let $\tau > 1$. There is a constant $c = c_\tau > 0$ such that, if V is any non-negative weight, and*

$$\frac{1}{|Q|} \int_Q V(x)\, (\log(e + V(x)/V_Q))^\tau\, dx \le c\, \ell(Q)^{-2}$$

for all cubes $Q \subset \mathbf{R}^3$, then

$$\int_{\mathbf{R}^3} |u(x)|^2\, V(x)\, dx \le \int_{\mathbf{R}^3} |\nabla u(x)|^2\, dx$$

for all $u \in \mathcal{C}_0^\infty(\mathbf{R}^3)$.

Exercises

8.1. Let ν be the positive Borel measure, defined on $\mathbf{R}^3$, given by

$$\nu(E) \equiv m_z(E \cap L_z),$$

where L_z is the z-axis and m_z is Lebesgue measure on L_z. Show that, for all cubes Q,

$$\frac{1}{|Q|}\nu(Q) \le c\ell(Q)^{-2}$$

for some $c > 0$. Nevertheless, show that there is no A such that

$$\int_{\mathbf{R}^3} |u(x)|^2 \, d\nu(x) \le A \int_{\mathbf{R}^3} |\nabla u(x)|^2 \, dx$$

holds for all $u \in \mathcal{C}_0^\infty(\mathbf{R}^3)$. (Hint: Write $x \in \mathbf{R}^3$ as $x = (x', z)$. Let $\psi \in \mathcal{C}_0^\infty(\mathbf{R}^3)$ be identically 1 for $|x| \le 1$. Let $\phi \in \mathcal{C}_0^\infty(\mathbf{R}^2)$ be identically 1 for $|x'| \le 1$ and identically 0 for $|x'| \ge 2$. Consider functions of the form

$$u(x) = \psi(x) \left(\sum_1^N \phi(2^k x') \right)$$

for large—but finite—N.) Then use a suitable approximation to show that 8.4 is not sufficient for 8.2.

8.2. Let ϕ be a fixed real, radial function in the Schwartz class on $\mathbf{R}^3$, chosen so that $\hat{\phi}(\xi)$ is always non-negative, and also satisfying

$$\hat{\phi}(\xi) = \begin{cases} 1 & \text{if } 3/4 < |\xi| < 4/3; \\ 0 & \text{if } |\xi| \le 1/2 \text{ or } |\xi| > 2. \end{cases}$$

Show that there exists a $\psi \in \mathcal{C}_0^\infty(\mathbf{R}^3)$ that is real, radial, supported in $\{x : |x| \le 1\}$, has integral equal to 0, and satisfies

$$\int_0^\infty \hat{\phi}(y\xi) \, \hat{\psi}(y\xi) \, \frac{dy}{y} \equiv 1$$

for all $\xi \ne 0$. (Hint: by convolving an appropriate function with itself, we can assume that $\hat{\psi}(\xi) \ge 0$ for all ξ.)

Notes

The results and basic approach in this chapter are taken from [62], and are based on those in [17] and [10]. There is nothing essential about our restriction to $\mathbf{R}^3$, and in fact Theorem 8.1 extends immediately to any $\mathbf{R}^d$ ($d > 2$) with no extra work. Theorem 8.1 has been significantly generalized by Pérez [48].

His result is as follows: If $d > 2$, then for all non-negative $V \in L^1_{\text{loc}}(\mathbf{R}^d)$ and all $f \in \mathcal{C}_0^\infty(\mathbf{R}^d)$,

$$\int_{\mathbf{R}^d} |f(x)|^2 \, V \, dx \leq C(d) \int_{\mathbf{R}^d} |\nabla f(x)|^2 \, N(V)(x) \, dx, \qquad (8.7)$$

where $N(\cdot)$ is a certain maximal function of V, which we now define. For $0 \leq \beta < d$, define

$$\mathcal{M}_\beta(V)(x) \equiv \sup_{\substack{x \in Q \\ Q \text{ a cube}}} \frac{1}{|Q|^{1-\beta/d}} \int_Q V(t) \, dt.$$

When $\beta = 0$ this is the Hardy-Littlewood maximal function $M(\cdot)$. When $\beta > 0$, $\mathcal{M}_\beta(\cdot)$ is naturally associated with the Riesz potential I_β,

$$I_\beta(f)(x) \equiv c_{\beta,d} \int_{\mathbf{R}^d} \frac{f(t)}{|x - t|^{d-\beta}} \, dt,$$

where $c_{\beta,d}$ is usually chosen so that

$$\widehat{I_\beta(f)}(\xi) = |2\pi\xi|^{-\beta} \hat{f}(\xi)$$

for all $f \in \mathcal{S}(\mathbf{R}^d)$. In [48] it is shown that 8.7 holds for $N(V)$ equal to $\mathcal{M}_2(M^2(V))$, where $M^2(\cdot)$ means a two-fold application of $M(\cdot)$. Pérez shows that the $M^2(\cdot)$ can be replaced by an Orlicz maximal operator $M_{B_\epsilon}(\cdot)$, where $B_\epsilon(x) \sim x(\log(e + x))^\epsilon$. (We discuss Orlicz maximal operators in chapter 10.) He also proves that

$$\int_{\mathbf{R}^d} |f(x)|^2 \, V \, dx \leq C(d) \int_{\mathbf{R}^d} |\Delta f(x)|^2 \, N(V)(x) \, dx$$

if $d > 4$ and $N(V) = M_4(M_{B_\epsilon}(V))$. His proofs do not use Littlewood-Paley theory, but rely on Orlicz maximal functions, Riesz potentials, and certain differential operators.

A different non-negativity criterion for the Schrödinger operator, based entirely on maximal functions of the form $\mathcal{M}_\beta$, was proved by Kerman and Sawyer in [32].

Finding a criterion for non-negativity is only the beginning of the analysis of the Schrödinger operator. The next step is to count the number of negative eigenvalues of $-\Delta - V$ in the cases when the operator is *not* non-negative. C. Fefferman and D. H. Phong [17] introduced the method of (essentially) identifying these negative eigenvalues with the minimal dyadic cubes Q where the non-negativity criterion fails. In our treatment, these would be the minimal dyadic Q such that

$$\frac{1}{|Q|} \int_Q V(x) \, (\log(e + V(x)/V_Q))^\tau \, dx > c \, \ell(Q)^{-2},$$

where c is a positive constant depending on d and τ. We have not explored this topic here. The interested reader can find treatments of it, for a variety of non-negativity criteria, in [17], [10], [32], and [11].

9

Some Singular Integrals

A *singular integral operator* is a type of linear operator defined by an integral that, strictly speaking, does not converge, but has to be evaluated in a principal value sense. What all of that means will become clear presently.

The most basic—and by far the most important—singular integral operator is the *Hilbert transform*, defined by the integral formula,

$$H(f)(x) \equiv \frac{1}{\pi} \int_{\mathbf{R}} \frac{f(y)}{x-y} \, dy. \tag{9.1}$$

Before proving anything about this operator, we should try to explain why anybody cares about operators like 9.1.

Formally, 9.1 is equal to convolution of f with $\frac{1}{\pi x}$, and so it should have a Fourier transform equal to

$$\hat{f}(\xi) \widehat{\frac{1}{\pi x}}(\xi).$$

Arguing formally (again), the Fourier transform of $\frac{1}{\pi x}$ is given by the principal value integral

$$\lim_{\epsilon \to 0^+} \frac{1}{\pi} \int_{\epsilon^{-1} > |x| > \epsilon} \frac{e^{-2\pi i x \xi}}{x} \, dx = \lim_{\epsilon \to 0^+} \frac{-i}{\pi} \int_{\epsilon < |x| < \epsilon^{-1}} \frac{\sin(2\pi x \xi)}{x} \, dx$$

$$= \begin{cases} -i & \text{if } \xi > 0; \\ i & \text{if } \xi < 0; \\ 0 & \text{if } \xi = 0. \end{cases}$$

In other words, assuming one accepts this formal reasoning,

$$\widehat{H(f)}(\xi) = -i\sigma(\xi)\hat{f}(\xi),$$

where $\sigma(\xi)$ is the familiar signum function:

$$\sigma(\xi) = \begin{cases} 1 & \text{if } \xi > 0; \\ -1 & \text{if } \xi < 0; \\ 0 & \text{if } \xi = 0. \end{cases}$$

Let us assume that f is real-valued. If we set $g(x) = (f(x) + iH(f)(x))/2$ then

$$\hat{g}(\xi) = \begin{cases} \hat{f}(\xi) & \text{if } \xi > 0; \\ 0 & \text{if } \xi < 0. \end{cases}$$

If g is regular enough, we can write:

$$g(x) = \int \hat{g}(\xi) e^{2\pi i x \xi} \, d\xi = \int_0^\infty \hat{f}(\xi) e^{2\pi i x \xi} \, d\xi. \tag{9.2}$$

The nice feature about 9.2 is that, if $\hat{f}$ is at all reasonable, then 9.2 even makes sense for $z = x + iy$, with $y > 0$. To wit, we can set

$$g(z) = \int_0^\infty \hat{f}(\xi) e^{2\pi i (x+iy)\xi} \, d\xi,$$

and the function $g(z)$ so defined is *analytic* on $\mathbf{R}_+^2$. It has real and imaginary parts u and v. As $y \to 0$, then (arguing formally!),

$$u(x,y) + iv(x,y) \to (f(x) + iH(f)(x))/2,$$

where we are leaving the precise mode of convergence unspecified.

Now, the function u is half of $P_y * f$, the Poisson integral of f which we briefly discussed in chapter 6. The function v, which is u's harmonic conjugate, is, plausibly, half of the Poisson integral of $H(f)$. We can thus outline the formal process that results in the operator 9.1: $f \mapsto u$ (by taking the Poisson integral); $u \mapsto v$ (integrate the Cauchy-Riemann equations); $v \mapsto H(f)$ (take v's boundary values).

Let's consider another example. If $f \in C_0^\infty(\mathbf{R}^d)$ and $d > 2$, then the differential equation

$$\Delta u = f$$
$$\lim_{|x| \to \infty} u(x) = 0,$$

has a unique solution given by the integral formula

$$u(x) = c_d \int \frac{f(y)}{|x-y|^{d-2}} \, dy,$$

where c_d is a constant. If we take two partial derivatives of u, then we get (arguing formally),

$$\frac{\partial^2 u}{\partial x_i \partial x_j} = \int f(y) \, \mathcal{K}_{i,j}(x-y) \, dy,$$

where

$$\mathcal{K}_{i,j}(x) = \begin{cases} c_d d(d-2) x_i x_j / |x|^{d+2} & \text{if } i \neq j; \\ c_d (2-d)(-dx_i^2 + \sum_l x_l^2)/|x|^{d+2} & \text{if } i = j. \end{cases} \tag{9.3}$$

The reader should notice a strong family resemblance between 9.1 and 9.3. First, both operators arise from similar processes: a differentiation (or an integration) is followed by an integration (or a differentiation), but these operations are not inverses of each other. Second, both operators are formally convolutions, in which the integrals very likely don't converge in the sense of Lebesgue. We can nevertheless use these convolutions to define bounded operators, because—point number three!—the kernels involved have some saving properties. These properties are captured in the following definition.

Definition 9.1. *A function* $\mathcal{K}$: $\mathbf{R}^d \setminus \{0\} \mapsto \mathbf{R}$ *is a* classical Calderón-Zygmund kernel *if: a) for all* $x \neq 0$,

$$|\mathcal{K}(x)| \leq |x|^{-d};$$

b) there is a positive $\alpha \leq 1$ *such that, for all* $x \neq 0$ *and all* $h \in \mathbf{R}^d$ *satisfying* $|h| \leq (1/2)|x|$,

$$|\mathcal{K}(x) - \mathcal{K}(x+h)| \leq \frac{|h|^\alpha}{|x|^{d+\alpha}};$$

c) for all $0 < r < R < \infty$,

$$\int_{x:\, r < |x| < R} \mathcal{K}(x)\, dx = 0.$$

Up to multiplication by positive constants, the convolution kernels for 9.1 and 9.3 are classical Calderón-Zygmund kernels (with $\alpha = 1$); we leave the proof of this as an exercise. If $\mathcal{K}$ is such a kernel, the *classical Calderón-Zygmund operator* $T_{\mathcal{K}}$ is defined by the integral

$$T_{\mathcal{K}}(f)(x) \equiv \int_{\mathbf{R}^d} \mathcal{K}(x - y)\, f(y)\, dy, \tag{9.4}$$

for functions f belonging to an appropriate test class (soon to be specified).

The integral 9.4 is likely to be undefined. We fix this problem by taking the integral in the principal value sense, i.e., by setting

$$\int_{\mathbf{R}^d} \mathcal{K}(x - y)\, f(y)\, dy = \lim_{\delta \to 0} \int_{\{y:\, |x-y| > \delta\}} \mathcal{K}(x - y)\, f(y)\, dy. \tag{9.5}$$

But it is still not obvious that this integral makes sense for "reasonable" f's (say, $f \in L^p$, for $1 \leq p < \infty$) even almost everywhere; and we are going to avoid this question.

Fortunately, it isn't hard to show that 9.5 makes sense for functions f that are fairly smooth and have bounded supports. Let f be supported in the ball $B(0; 1)$ and have two continuous derivatives. I claim that $T_{\mathcal{K}}(f)$ is not only defined for all x, but bounded, and Hölder continuous of order 1.

The "defined" part is easy. Let $0 < \eta_1 < \eta_2$ be two very small numbers. Then:

$$\left| \int_{\eta_1 < |x-y| < \eta_2} \mathcal{K}(x - y) \, f(y) \, dy \right| \leq C\eta_2, \tag{9.6}$$

where C is a constant that only depends on d and the supremum of $|\nabla f|$. The proof of 9.6 is simple—once you know the trick. Let B be $|\nabla f|$'s supremum. Since

$$\int_{\eta_1 < |x-y| < \eta_2} \mathcal{K}(x - y) \, dy = 0,$$

we have

$$\left| \int_{\eta_1 < |x-y| < \eta_2} \mathcal{K}(x - y) \, f(y) \, dy \right| = \left| \int_{\eta_1 < |x-y| < \eta_2} \mathcal{K}(x - y) \, (f(y) - f(x)) \, dy \right|$$
$$\leq B \int_{\eta_1 < |x-y| < \eta_2} \frac{|x - y|}{|x - y|^d} \, dy$$
$$\leq CB\eta_2.$$

Inequality 9.6 easily implies the existence of the principal value integral 9.5. As for smoothness, we note that, for any x and x',

$$T_\mathcal{K}(f)(x) - T_\mathcal{K}(f)(x') = \int_{\mathbf{R}^d} \mathcal{K}(x - y) \, (f(y) - f(y + x' - x)) \, dy, \tag{9.7}$$

because the operator $T_\mathcal{K}$ commutes with translations. But, as functions of y, the first partial derivatives of $f(y) - f(y + x' - x)$ have their sizes bounded by a constant times $|x - x'|$. Now a repetition of the earlier argument implies $|T_\mathcal{K}(f)(x) - T_\mathcal{K}(f)(x')| \leq C|x - x'|$. The form of $T_\mathcal{K}(f)$ shows that $T_\mathcal{K}(f)(x) \to 0$ fairly fast as $|x| \to \infty$, implying that $T_\mathcal{K}(f)$ is a bounded function. In fact, because of the decay in $\mathcal{K}$, we have

$$|T_\mathcal{K}(f)(x)| \leq C(1 + |x|)^{-d} \tag{9.8}$$

for large x. Since $T_\mathcal{K}(f)$ is bounded, 9.8 holds for all x (with a possibly larger C). It is clear that similar arguments will imply similar estimates if $B(0; 1)$ is replaced by any other ball. Therefore, $T_\mathcal{K}(f)$ is defined on a dense subspace of $L^p(\mathbf{R}^d)$ when $1 < p < \infty$. The problem now is to prove that $T_\mathcal{K}$ satisfies an L^p bound on this subspace; in other words, that

$$\|T_\mathcal{K}(f)\|_p \leq C\|f\|_p \tag{9.9}$$

for all $f \in \mathcal{C}_0^\infty(\mathbf{R}^d)$, with a constant C independent of f. Inequality 9.9 will then imply that $T_\mathcal{K}$ has a unique, bounded extension to all of L^p. Our work on the square function will make proving 9.9 a fairly easy task. Since we already know that $T_\mathcal{K}(f) \in L^p$, it is enough to show that

$$S_\psi(T_{\mathcal{K}}(f))(x) \le CG_\beta(f)(x) \tag{9.10}$$

pointwise, for some fixed $\beta > 0$, where $S_\psi(\cdot)$ and $G_\beta(\cdot)$ are (respectively) the real-variable and intrinsic square functions we defined in chapter 6. (See 6.1 and the sentence immediately following, and see Definition 6.1.)

Let $\psi \in \mathcal{C}_0^\infty(\mathbf{R}^d)$ be real, radial, be supported in $\{x : |x| \le 1\}$, have integral equal to 0, and be normalized so that, for all $\xi \in \mathbf{R}^d \setminus \{0\}$,

$$\int_0^\infty |\hat{\psi}(y\xi)|^2 \frac{dy}{y} = 1.$$

(Formula 5.1 showing up again.) Inequality 9.10 will follow if we can show that

$$|(T_{\mathcal{K}}(f)) * \psi_y(t)| \le C\tilde{A}_{(\beta,\epsilon)}(f)(t,y), \tag{9.11}$$

for all $(t,y) \in \mathbf{R}_+^{d+1}$, for some fixed, positive β and ϵ.

Inequality 9.11 will follow from the equation

$$(T_{\mathcal{K}}(f)) * \psi_y(t) = f * (T_{\mathcal{K}}(\psi_y))(t) \tag{9.12}$$

and some estimates on $T_{\mathcal{K}}(\psi_y)(t)$. The equation is easy. For every $\delta > 0$, set

$$T_{\mathcal{K},\delta}(f)(x) \equiv \int_{\{y:\ |y|>\delta\}} f(x-y)\mathcal{K}(y)\,dy.$$

The previous arguments imply that if $f \in \mathcal{C}_0^\infty(\mathbf{R}^d)$, then $T_{\mathcal{K},\delta}(f) \to T_{\mathcal{K}}(f)$ uniformly and in L^p when $1 < p < \infty$. (The convergence in L^p holds because, when x is large, $T_{\mathcal{K},\delta}(f)(x) = T_{\mathcal{K}}(f)(x)$ when δ is less than the distance to f's support.) It is therefore trivial that

$$(T_{\mathcal{K},\delta}(f)) * \psi_y(t) = f * (T_{\mathcal{K},\delta}(\psi_y))(t) \tag{9.13}$$

for all $\delta > 0$, and that both sides of 9.13 converge to both sides of 9.12 as $\delta \to 0$. That gives us 9.12.

The estimates are also not too bad. The essential step is to show that $T_{\mathcal{K}}(\psi)$ is a positive multiple of a function in $\mathcal{C}_{(1,\alpha)}$. An additional argument will show that, for every $y > 0$, $T_{\mathcal{K}}(\psi_y)$ is a boundedly positive multiple of a function of the form $\phi_y^{(y)}$, where $\phi^{(y)} \in \mathcal{C}_{(1,\alpha)}$. This function $\phi^{(y)}$ will in general *not* be $T_{\mathcal{K}}(\psi)$, but will—as we have indicated—vary with y. But that will be sufficient to yield 9.11.

We almost have what we need. Our earlier arguments show that

$$|T_{\mathcal{K}}(\psi)(x) - T_{\mathcal{K}}(\psi)(x')| \le C|x-x'|$$

holds for any x and x'. We also know that

$$|T_{\mathcal{K}}(\psi)(x)| \le C(1+|x|)^{-d}.$$

We need to show that $T_{\mathcal{K}}(\psi)$ and its Hölder modulus actually decay a little faster than this, and that $\int_{\mathbf{R}^d} T_{\mathcal{K}}(\psi)\, dx = 0$.

Size decay estimate. It is clear that, if x and x' both have moduli larger than 3, and $|x - x'| \leq 1$, then

$$|\mathcal{K}(x) - \mathcal{K}(x')| \leq C(d)|x - x'|^{\alpha}|x|^{-d-\alpha}.$$

Now, when $|x| \geq 3$,

$$|T_{\mathcal{K}}(\psi)(x)| = \left| \int \mathcal{K}(x - y)\, \psi(y)\, dy \right|$$

$$= \left| \int (\mathcal{K}(x - y) - \mathcal{K}(x))\, \psi(y)\, dy \right|,$$

because $\int \psi\, dy = 0$. Because of the smoothness bound on K, and the fact that ψ is supported in the unit ball, the last integral is no bigger than a constant times $|x|^{-d-\alpha}$. This shows that $T_{\mathcal{K}}(\psi)(x)$ decays fast enough for us.

Smoothness decay estimate. Once again, we can assume that x and x' have moduli larger than 3 and $|x - x'| \leq 1$. We write

$$|T_{\mathcal{K}}(\psi)(x) - T_{\mathcal{K}}(\psi)(x')| = \left| \int (\mathcal{K}(x - y) - \mathcal{K}(x' - y))\, \psi(y)\, dy \right|$$

$$= \left| \int \mathcal{K}(x - y)\, (\psi(y) - \psi(y + x' - x))\, dy \right|$$

$$= \left| \int (\mathcal{K}(x - y) - \mathcal{K}(x))\, (\psi(y) - \psi(y + x' - x))\, dy \right|$$

$$\leq C|x|^{-d-\alpha}|x - x'|$$

$$\leq C|x - x'|(1 + |x|)^{-d-\alpha},$$

which is just what we needed.

Cancelation. Recall that, for $\delta > 0$, we defined

$$T_{\mathcal{K},\delta}(\psi)(x) \equiv \int_{\{y:\ |y|>\delta\}} \psi(x - y)\, \mathcal{K}(y)\, dy.$$

Our decay estimates imply that this is an L^1 function for every $\delta > 0$. As $\delta \to 0$, this function converges in the L^1 norm to the L^1 function $T_{\mathcal{K}}(\psi)$. (The reason the convergence is in L^1, and not merely uniform, is that, when $|x| \geq 2$, $T_{\mathcal{K}}(\psi)(x) = T_{\mathcal{K},\delta}(\psi)(x)$ as soon as $\delta < 1$.) Thus, it is enough to show that $\int T_{\mathcal{K},\delta}(\psi)\, dx = 0$ for every $\delta > 0$. Now fix $\delta > 0$ and, for $R > \delta$, define

$$T_{\mathcal{K},\delta,R}(\psi)(x) \equiv \int_{\{y:\ R>|y|>\delta\}} \psi(x - y)\, \mathcal{K}(y)\, dy.$$

The function $T_{\mathcal{K},\delta,R}(\psi)$ is the convolution of ψ with an L^1 function, and therefore

$$\int T_{\mathcal{K},\delta,R}(\psi)\,dx = 0$$

for all R (it inherits cancelation from ψ). Let us set

$$T_{\mathcal{K},R}(\psi)(x) \equiv \int_{|y|>R} \psi(x-y)\,\mathcal{K}(y)\,dy,$$

and notice that $T_{\mathcal{K},\delta}(\psi) = T_{\mathcal{K},\delta,R}(\psi) + T_{\mathcal{K},R}(\psi)$. We will be done if we can show

$$\lim_{R\to\infty} \int T_{\mathcal{K},R}(\psi)(x)\,dx = 0.$$

This integral is easy to estimate. If $|x| < R - 1$, $T_{\mathcal{K},R}(\psi)(x) = 0$. If $R - 1 \le |x| \le R + 1$, $|T_{\mathcal{K},R}(\psi)(x)|$ is bounded by a constant times $|x|^{-d}$. If $|x| > R + 1$, $|T_{\mathcal{K},R}(\psi)(x)|$ is less than or equal to a constant times $|x|^{-d-\alpha}$. Putting everything together,

$$\int T_{\mathcal{K},R}(\psi)(x)\,dx$$

is seen to be no bigger than a constant times $R^{-1} + R^{-\alpha}$, and that is just what we needed.

Now we want to show that $T_{\mathcal{K}}(\psi_y)$ is, up to a bounded positive multiple, a function of the form $\phi_y^{(y)}$, where $\phi^{(y)} \in \mathcal{C}_{(1,\alpha)}$. There are at least two ways to do this. The longer way is to repeat the preceding argument, making allowance (in several places) for the dilation by y. The shorter way is to observe that $T_{\mathcal{K}}(\psi_y) = (T_{\tilde{\mathcal{K}}}(\psi))_y$, where $\tilde{\mathcal{K}}(x) = y^d \mathcal{K}(yx)$; and to note that $\tilde{\mathcal{K}}$ is also a classical Calderón-Zygmund kernel. Thus, $T_{\mathcal{K}}(\psi_y)$ will be a boundedly positive multiple of some $\phi_y^{(y)}$, where $\phi^{(y)} \in \mathcal{C}_{(1,\alpha)}(\mathbf{R}^d)$. From this we immediately get inequality 9.11 (with $\beta = 1$ and $\epsilon = \alpha$), which in turn yields inequality 9.10 for any β strictly less than α.

We have now shown that $T_{\mathcal{K}}$ is well-defined as a bounded linear operator on L^p. But this definition is a little troublesome, because we obtained it as an extension from $\mathcal{C}_0^\infty(\mathbf{R}^d)$. It would be nice to have a tractable definition of $T_{\mathcal{K}}(f)$ that worked directly with f. By "tractable" we mean one that avoids directly evaluating the singular integral 9.4. Our square function machinery gives us a straightforward, fairly constructive way to do this.

Take another look at 9.12. We know it holds for $f \in \mathcal{C}_0^\infty(\mathbf{R}^d)$. However, both sides depend continuously on f, with respect to the L^p norm. Therefore, 9.12 holds for *all* $f \in L^p$. The reader should spend a minute thinking of what this means. The right-hand side of 9.12 is the convolution of f with a dilate of a function in $\mathcal{C}_{(1,\alpha)}$: it is something we can know *as a function*. The left-hand side of 9.12 is convolution of $T_{\mathcal{K}}(f)$ with ψ_y, where $T_{\mathcal{K}}(f)$ is only "known" as a vector in L^p. Therefore, we can use 9.12 to understand $T_{\mathcal{K}}(f)$ as a function.

Let K have compact closure contained in $\mathbf{R}_+^{d+1}$. If $f \in L^p$, $1 < p < \infty$, then the following integral makes sense, and defines a function in $\mathcal{C}_0^\infty(\mathbf{R}^d)$:

$$P_K(f)(x) \equiv \int_K (f * (T_{\mathcal{K}}(\psi_y)))(t)\, \psi_y(x-t)\, \frac{dt\, dy}{y}. \tag{9.14}$$

By what we said in the preceding paragraph, $P_K(f)$ is equal to

$$\int_K ((T_{\mathcal{K}}(f) * \psi_y(t))\, \psi(x-t)\, \frac{dt\, dy}{y}. \tag{9.15}$$

Let $K_1 \subset K_2 \subset \cdots$ be a compact-measurable exhaustion of $\mathbf{R}_+^{d+1}$. Our work with the Calderón reproducing formula now implies that $P_{K_i}(f) \to T_{\mathcal{K}}(f)$ in L^p. But this is a sequence of integrals of the form

$$\int_{K_i} (f * \Phi_y^{(y)}(t))\, \psi_y(x-t)\, \frac{dt\, dy}{y},$$

where $\Phi^{(y)} = T_{\bar{\mathcal{K}}}(\psi)$. If we denote the L^p limit of these integrals by

$$\int_{\mathbf{R}_+^{d+1}} (f * \Phi_y^{(y)}(t))\, \psi_y(x-t)\, \frac{dt\, dy}{y}, \tag{9.16}$$

then 9.16 can be our definition of $T_{\mathcal{K}}$ on L^p.

The bound 9.10 and our earlier work on square functions lead quickly to weighted norm inequalities for the singular integral operators $T_{\mathcal{K}}$. Recall the inequality from chapter 4:

$$\tilde{S}_{sd}(f)(x) \leq C S_{\psi,\alpha}(f)(x),$$

valid for some sufficiently large $\alpha > 0$. Therefore, if $f \in \mathcal{C}_0^\infty(\mathbf{R}^d)$,

$$\tilde{S}_{sd}(T_{\mathcal{K}}(f))(x) \leq C S_{\psi,\alpha}(T_{\mathcal{K}}(f))(x) \leq C G_\tau(f)(x),$$

for appropriate $\tau > 0$. If $1 < p < \infty$, and if v and w are weights such that

$$\int_Q v(x)\, (\log(e + v(x)/v_Q))^r\, dx \leq \int_Q w(x)\, dx$$

for all cubes Q, for some $r > p/2$, then

$$\int |T_{\mathcal{K}}(f)|^p\, v\, dx \leq C \int (\tilde{S}_{sd}(T_{\mathcal{K}}(f)))^p\, w\, dx, \tag{9.17}$$

implying

$$\int |T_{\mathcal{K}}(f)|^p\, v\, dx \leq C \int (G_\tau(f))^p\, w\, dx.$$

If $1 < p \leq 2$, then

$$\int (G_\tau(f))^p \, w \, dx \leq C \int |f|^p \, M(w) \, dx.$$

Inequality 9.17 holds with $w = M(v)$ for $1 < p < 2$ and with $w = M^2(v)$ when $p = 2$. Therefore, if $f \in \mathcal{C}_0^\infty(\mathbf{R}^d)$ and $1 < p \leq 2$,

$$\int |T_\mathcal{K}(f)|^p \, v \, dx \leq C \int |f|^p \, M^*(v) \, dx \qquad (9.18)$$

for all weights v, where we may take

$$M^*(v) = \begin{cases} M^2(v) & \text{if } 1 < p < 2; \\ M^3(v) & \text{if } p = 2. \end{cases}$$

In the next chapter we will learn how to generalize inequalities of the form 9.18 to all $1 < p < \infty$.

Exercises

9.1. For $d > 1$, define S^{d-1} to be the set $\{x \in \mathbf{R}^d : |x| = 1\}$, the boundary of $\{x : |x| \leq 1\}$. Let $\Omega : S^{d-1} \mapsto \mathbf{R}$ be Hölder continuous of order $\alpha > 0$, and suppose that

$$\int_{S^{d-1}} \Omega(x) \, d\sigma(x) = 0,$$

where $d\sigma$ denotes Euclidean surface measure. Show that, up to multiplication by a positive constant, the formula

$$\mathcal{K}(x) = \frac{\Omega(x/|x|)}{|x|^d}$$

defines a classical Calderón-Zygmund kernel. Show that, if $\psi \in \mathcal{C}_0^\infty(\mathbf{R}^d)$, then, for all $y > 0$,

$$T_\mathcal{K}(\psi_y) \equiv (T_\mathcal{K}(\psi))_y,$$

implying that, if $f \in L^p$ $(1 < p < \infty)$, we can write

$$T_\mathcal{K}(f)(x) = c \int_{\mathbf{R}_+^{d+1}} (f * \Phi_y(t)) \, \psi_y(x - t) \, \frac{dt \, dy}{y}$$

for a suitable $\psi \in \mathcal{C}_0^\infty(\mathbf{R}^d)$, constant c, and $\Phi \in \mathcal{C}_{(1,\alpha)}(\mathbf{R}^d)$.

9.2. Let $w \in A_p$ for $1 < p < \infty$, and let $T_\mathcal{K}$ be a classical Calderón-Zygmund operator. Show that

$$\int_{\mathbf{R}^d} |T_\mathcal{K}(f)(x)|^p \, w \, dx \leq C(p, w) \int_{\mathbf{R}^d} |f(x)|^p \, w \, dx$$

for all $f \in \mathcal{C}_0^\infty(\mathbf{R}^d)$. Use the results from exercise 7.10 to show how one can extend $T_\mathcal{K}$ to a bounded operator on all of $L^p(w)$. Show that if $f \in L^p(w)$, and $\{K_i\}$ is any compact-measurable exhaustion of $\mathbf{R}_+^{d+1}$, then $T_\mathcal{K}(f)$ equals the $L^p(w)$ limit of $P_{K_i}(f)$, where $P_K(f)$ is as defined by 9.14 and 9.15.

9.3. Let $1 < p < \infty$ and let $w \in L_{\text{loc}}^1(\mathbf{R}^d)$ be a weight. Suppose that every classical Calderón-Zygmund operator $T_\mathcal{K}$ defines a bounded operator mapping $L^p(w)$ into $L^p(w)$. Show that $w \in A_p$. (Hint: First show that w is doubling. Then see exercise 7.7 for an idea of what sort of function to test the boundedness criterion against.)

Notes

The $L^p(w)$ boundedness ($1 < p < \infty$, $w \in A_p$) was proved for the Hilbert transform by Hunt, Muckenhoupt, and Wheeden in [30], where they also proved the necessity of the A_p condition. Their result was generalized to general classical Calderón-Zygmund operators $T_\mathcal{K}$ by Coifman and Fefferman in [13]. Inequalities of the form

$$\int |T_\mathcal{K}(f)|^p \, v \, dx \le C \int |f|^p \, \tilde{M}(v) \, dx,$$

with $\tilde{M}$ being some maximal function, first appeared in [13], with $\tilde{M}(v)$ being $(M(v^r))^{1/r}$ for any $r > 1$. Inequalities in which $\tilde{M}$ is some finite iterate of M were first proved in [62], but only for $1 < p \le 2$. By a clever use of Orlicz maximal functions, Pérez [46] was able to reprove these results and extend them to large p *without* using Littlewood-Paley theory. The method we have followed, of defining the action of $T_\mathcal{K}$ via the Calderón reproducing formula, is similar to that in [62]; see also [22], where the this is done in the context of Besov spaces.

Orlicz Spaces

Many precise results in analysis call for a scale of integrability conditions more flexible than that provided by the L^p spaces. We have encountered such a scale in the two-weight norm inequalities studied in chapter 3 and chapter 4. But, more simply, we could consider the example given by

$$f(x) \equiv \begin{cases} |x| \ln|x||^r|^{-1} & \text{if } |x| \leq 1/2; \\ 0 & \text{otherwise,} \end{cases}$$

where $r > 1$. The function f belongs to L^1, but not to L^p for any $p > 1$. However, this is not the whole story. If $|x|$ is small then $f(x)|\ln(f(x))|^s$ is comparable to $|x| \ln|x||^{r-s}|^{-1}$, and so $f|\ln f|^s$ will still be in L^1 if $0 < s < r-1$. In other words, when $r > 1$, this function is in L^1—and a little more—without being in L^p for any higher p.

A useful, finely structured scale of integrability conditions is provided by the theory of Orlicz spaces, which are natural generalizations of the L^p spaces—in the following sense. Consider: we can say that $f \in L^p$ if and only if $\Phi(|f|) \in L^1$, where $\Phi(t) = t^p$. The theory of Orlicz spaces applies this idea to more general functions Φ.

We say that $\Phi : [0, \infty] \mapsto [0, \infty]$ is a *Young function* if:
a) $\Phi(0) = 0$;
b) there exists a number $0 < x_0 < \infty$ such that $\Phi(x_0) < \infty$;
c) $\Phi \nearrow \infty$; i.e., Φ is increasing and $\lim_{x \to \infty} \Phi(x) = \infty$;
d) Φ is convex, where this means that, for all $0 \leq x \leq y \leq \infty$, and all $0 \leq t \leq 1$, $\Phi(tx + (1 - t)y) \leq t\Phi(x) + (1 - t)\Phi(y)$.

Before going on, let's observe some features and easy consequences of this definition.

1. We allow Φ to take on the value $+\infty$.
2. We do not require Φ to be either *strictly* convex or *strictly* increasing; however, unless Φ takes on the value $+\infty$ for some finite x_0, $\Phi(x)$ must be strictly increasing for sufficiently large x.
3. $\Phi(\infty) = \infty$.

4. Either Φ is continuous on all of $[0, \infty)$ or there is a $0 < b < \infty$ such that Φ is continuous on $[0, b)$ and identically infinite on $(b, \infty]$. In particular, Φ is continuous at 0.

Now let's suppose we are working on a measure space $(X, \mathcal{M}, \mu)$. Given a Young function Φ, we define the *Orlicz space* $L^\Phi(X, \mathcal{M}, \mu)$ to be the set of measurable f such that

$$\int \Phi(|f|/\lambda)\, d\mu < \infty$$

for some $\lambda > 0$. We put a norm on L^Φ by defining

$$\|f\|_\Phi = \inf\{\lambda > 0 : \int \Phi(|f|/\lambda)\, d\mu \leq 1\}.$$

It will be helpful to observe some features of this so-called norm.

1. It really is a norm. It's trivial to see that $\|\alpha f\|_\Phi = |\alpha|\|f\|_\Phi$ and that $\|f\|_\Phi = 0$ if and only if $f = 0$ μ-a.e. The triangle inequality is only slightly trickier. Let f and g be non-negative functions in L^Φ, with respective norms λ_1 and λ_2, both of which we may take to be positive. We need to show that $\|f + g\|_\Phi \leq \lambda_1 + \lambda_2$. Let $\lambda_1 < \gamma_1$ and $\lambda_2 < \gamma_2$. Then

$$\int \Phi(f/\gamma_1)\, d\mu$$

and

$$\int \Phi(g/\gamma_2)\, d\mu$$

are both ≤ 1. We will be done if we can show that

$$\int \Phi((f + g)/(\gamma_1 + \gamma_2))\, d\mu \leq 1.$$

Fortunately, this is easy. By simple algebra:

$$\frac{f + g}{\gamma_1 + \gamma_2} = \frac{f}{\gamma_1}\frac{\gamma_1}{\gamma_1 + \gamma_2} + \frac{g}{\gamma_2}\frac{\gamma_2}{\gamma_1 + \gamma_2}.$$

The convexity of Φ now implies that

$$\int \Phi((f + g)/(\gamma_1 + \gamma_2))\, d\mu \leq \frac{\gamma_1}{\gamma_1 + \gamma_2}\int \Phi(f/\gamma_1)\, d\mu + \frac{\gamma_2}{\gamma_1 + \gamma_2}\int \Phi(g/\gamma_2)\, d\mu$$

$$\leq \frac{\gamma_1}{\gamma_1 + \gamma_2} + \frac{\gamma_2}{\gamma_1 + \gamma_2} = 1,$$

proving the result. Armed with this fact, the reader should have little trouble showing that L^Φ is complete, and we urge him to do it.

2. If Φ_1 and Φ_2 are two Young functions, and $\Phi_1(x) \leq \Phi_2(x)$ for all x, then $\|f\|_{\Phi_1} \leq \|f\|_{\Phi_2}$. Therefore if there are positive constants a and b such that

$\Phi_1(ax) \le \Phi_2(x) \le \Phi_1(bx)$ for all x, then $L^{\Phi_1} = L^{\Phi_2}$, with comparability of norms.

3. If Φ_1 and Φ_2 are two Young functions, then $\Phi \equiv \Phi_1 + \Phi_2$ is a Young function, and $L^\Phi = L^{\Phi_1} \cap L^{\Phi_2}$.

Let's look at some examples.

1. If $\Phi(x) = x^p$ $(1 \le p < \infty)$, then $L^\Phi = L^p(X, \mathcal{M}, \mu)$ and $\|f\|_\Phi = \|f\|_p$.

2. If we let $p \to \infty$ in the previous example, we get

$$\Phi(x) = \begin{cases} 0 & \text{if } 0 \le x < 1; \\ 1 & \text{if } x = 1; \\ \infty & \text{otherwise.} \end{cases}$$

This is a Young function according to our definition. The corresponding Orlicz space L^Φ is L^∞, with the usual norm.

3. Now suppose we take

$$\Phi(x) = \begin{cases} x/(1-x) & \text{if } 0 \le x < 1; \\ \infty & \text{otherwise.} \end{cases}$$

The reader should check that Φ is a Young function. If our measure space is finite, this Φ also yields $L^\Phi = L^\infty$, though not with equality of norms, even if we also assume $\mu(X) = 1$. To see that the norms aren't equal, take f identically equal to 1 and suppose that $\int \Phi(f/\lambda) \, d\mu \le 1$. Then $\lambda > 1$, and the integral is equal to $1/(\lambda - 1)$, implying that $\lambda \ge 2$. If the measure space is infinite, we do not get $L^\Phi = L^\infty$, but $L^\Phi = L^1 \cap L^\infty$: since $\Phi(x)$ is essentially x near 0, the functions in L^Φ must not only be bounded, but must (in an average sense) decay rapidly "at infinity."

4. Define $\Phi(x) = \exp(x^2) - 1$. If the measure space is finite, L^Φ consists of the exponentially square-integrable functions: $f \in L^\Phi$ if and only if there is an $\epsilon > 0$ such that

$$\int \exp(\epsilon |f|^2) \, d\mu < \infty.$$

(The -1 in the defintion of Φ is unimportant here.) If the space is infinite then L^Φ consists of all of the L^2 functions f such that, for some $\epsilon > 0$,

$$\int_{\{x: \, |f(x)| > 1\}} \exp(\epsilon |f|^2) \, d\mu < \infty.$$

The restriction to L^2 comes because, when f is small, $\Phi(|f|) \sim |f|^2$. This restriction is similar to the restriction to L^1 in example 3.

5. Consider the two functions $\Phi_1(x) = x \log(1 + x)$ and $\Phi_2(x) = x \log(2 + x)$. If the measure space is finite then $L^{\Phi_1} = L^{\Phi_2}$, with comparable but different norms; the common space is called $L \log L$. But if the measure space is $\mathbf{R}$ with Lebesgue measure, the spaces will not be the same. We urge the reader to find an f in this setting that lies in one space but not in the other.

6. Let $1 \leq r < \infty$ and $s \in \mathbf{R}$. If $s \geq 0$, then $x^r(\log(e+x))^s$ is a Young function for any $r \geq 1$. But what if s is negative? We notice that $x^r(\log(e+x))^s$ is *eventually* increasing. However, it will not be convex—even eventually—unless $r > 1$. So, if $s < 0$, we will *require* that $r > 1$. In that case we can define a Young function Φ that is linear for "small" x and that equals a constant multiple of $x^r(\log(1+x))^s$ when x is "large." When $s > 0$, the L^Φ norm is marginally stronger than the L^r norm, but, when the measure space is finite, it is weaker than the $L^{r+\epsilon}$ norm, for any $\epsilon > 0$. If $s < 0$—and $r > 1$—L^Φ's norm is weaker than L^r's, but (in the case of finite total measure) stronger than all the $L^{r-\epsilon}$ norms. We will soon be taking a closer look at these spaces.

Since L^Φ is a Banach space (do the exercise!), it has a dual space. Sometimes this dual space is L^Ψ for some Ψ, and sometimes it isn't. The functions $\Phi = x^p$ $(1 \leq p < \infty)$ provide examples of the first type. If $1 < p$ then we can take $\Psi = x^{p'}$, where $p' = p/(p-1)$, p's dual exponent. If $p = 1$ then we can take as Ψ the Young function which gave us L^∞ earlier, or we can take a shortcut and set

$$\Psi(x) = \begin{cases} 0 & \text{if } 0 \leq x \leq 1; \\ \infty & \text{otherwise.} \end{cases}$$

(The reader should check that the Orlicz space defined by this Ψ is also L^∞.) However, if Φ equals the Ψ we just wrote, which makes L^Φ equal to L^∞, its dual is not an Orlicz space except in trivial cases. This suggests two questions: a) Given a Young function Φ, when is $(L^\Phi)^*$ an Orlicz space? b) If $(L^\Phi)^*$ equals some L^Ψ, what is Ψ? Fortunately, we won't need to know the answers to either a) or b) in general. But we will need to know this: Given a Young function Φ, do there exist a constant C and a Young function Ψ such that

$$\int |fg|\, d\mu \leq C\|f\|_\Phi \|g\|_\Psi$$

for all measurable f and g; and, if so, what is a "good" or "natural" Ψ?

What we need is an Orlicz space version of Hölder's inequality.

Definition 10.1. *If Φ is a Young function, its dual function, $\bar{\Phi} : [0, \infty] \mapsto [0, \infty]$ is defined by:*

$$\bar{\Phi}(y) = \sup\{xy - \Phi(x) : \ x \geq 0\}.$$

Remark. This definition looks peculiar. The main reason for its strange form is to ensure

$$xy \leq \Phi(x) + \bar{\Phi}(y)$$

for all x and y. This numerical inequality is the heart of the proof of the general Hölder inequality.

The reader might want to come back to the next theorem after looking at the examples which follow it.

Theorem 10.1. *If Φ is a Young function, so is $\bar{\Phi}$.*

Proof of Theorem 10.1. For any y, $xy - \Phi(x)$ equals 0 when $x = 0$, thus $\bar{\Phi}(y) \geq 0$; and it is trivial that $\bar{\Phi}(0) = 0$. If $0 \leq y_1 \leq y_2$ then, for any x, $xy_2 - \Phi(x) \geq xy_1 - \Phi(x)$, implying that $\bar{\Phi}(y_2) \geq \bar{\Phi}(y_1)$, and that $\bar{\Phi} \nearrow$. By definition, there is a number $0 < x_0 < \infty$ such that $\Phi(x_0) < \infty$. Thus, for every y, $\bar{\Phi}(y) \geq x_0 y - \Phi(x_0)$, which goes to ∞ as $y \to \infty$. Since Φ is convex and $\Phi(0) = 0$, the function $\Phi(x)/x$ increases to a positive (possibly infinite) value as $x \to \infty$. Call this limit m and suppose that $0 < y < m$. Then $xy - \Phi(x) \to -\infty$ as $x \to \infty$, implying that $\bar{\Phi}(y)$ is finite. (N.B. It might be 0.) The only thing left to prove about $\bar{\Phi}$ is its convexity. If $0 \leq y_1 \leq y_2 < \infty$ and $0 < t < 1$ then

$$x(ty_1 + (1 - t)y_2) - \Phi(x) = t(xy_1 - \Phi(x)) + (1 - t)(xy_2 - \Phi(x))$$
$$\leq t\bar{\Phi}(y_1) + (1 - t)\bar{\Phi}(y_2)$$

for all $x \geq 0$. Taking the supremum on the left-hand side yields $\bar{\Phi}(ty_1 + (1 - t) y_2) \leq t\bar{\Phi}(y_1) + (1 - t)\bar{\Phi}(y_2)$; easy arguments (which we invite the reader to do) take care of the endpoint cases $t = 0$, $t = 1$, and $y_2 = \infty$. That was the last piece: $\bar{\Phi}$ is a Young function.

Now some examples (which we encourage the reader to work out).

1. Let $\Phi(x) = x^r/r$, where $1 < r < \infty$. This makes L^Φ equal to L^r, but with a slightly different norm. The supremum of $xy - \Phi(x)$ is attained when $x^{r-1} = y$, and it equals $y^{r'}/r'$, where r' is r's dual exponent.

2. Let $\Phi(x) = x$, making $L^\Phi = L^1$. If $y \leq 1$ then the supremum of $xy - \Phi(x)$ is 0 (attained when $x = 0$). If $y > 1$ its supremum is infinite. Thus, the $\bar{\Phi}$ we obtain is one whose $L^{\bar{\Phi}}$ is L^∞.

3. Now define

$$\Phi(x) = \begin{cases} 0 & \text{if } 0 \leq x \leq 1; \\ \infty & \text{if } x > 1, \end{cases}$$

and consider $xy - \Phi(x)$. This equals xy for $0 \leq x \leq 1$ and $-\infty$ when $x > 1$. Therefore its supremum—$\bar{\Phi}(y)$—is just y.

4. Let $\Phi(x) = \exp(x^2) - 1$. Then $xy - \Phi(x)$ attains its supremum when $y = 2x \exp(x^2)$. If y is *large*, this happens when $x \sim (\log y)^{1/2}$, and $xy - \Phi(x) = (2x^2 - 1)\exp(x^2) + 1 \sim y(\log y)^{1/2}$. In this case we don't get a precise formula but, what is just as useful, an estimate of $\bar{\Phi}(y)$ for large y: $\bar{\Phi}(y) \sim y(\log y)^{1/2}$.

5. If, in the previous example, we had taken $\Phi(x) = \exp(x^r) - 1$, with $r \geq 1$, we would have got $\bar{\Phi}(y) \sim y(\log y)^{1/r}$ for large y.

6. If $0 < r < 1$ then $\exp(x^r) - 1$ is convex for large x. We can build a Young function $\Phi(x)$ that equals $\exp(x^r) - 1$ when x is big. Repeating the construction from example 4, we get $\bar{\Phi}(y) \sim y(\log y)^{1/r}$ for large y.

7. Let $\Phi(x) = x(\log(e+x))^\alpha$, where $\alpha > 0$. If y is large, the quantity $xy - \Phi(x)$ is maximized when $y \sim (\log(e + x))^\alpha$, or $x \sim \exp(c'_\alpha y^{1/\alpha})$. This implies that, for large y, $\bar{\Phi}(y) \sim \exp(c_\alpha y^{1/\alpha})$ for some positive constant c'_α.

8. Let $\Phi(x) \sim (x(\log(1 + x))^s)^r$ for large x, where $r > 1$ and $s \in \mathbf{R}$. If y is large, we maximize $xy - \Phi(x)$ by setting $x \sim (y/(\log(1 + y))^{sr})^{1/(r-1)}$. Then $\bar{\Phi}(y) \sim xy - \Phi(x) \sim y^{r'}(\log(1 + y))^{-sr/(r-1)} = (y(\log(1 + y))^{-s})^{r'}$.

The point of the dual Young function is this:

Theorem 10.2. *Let Φ be a Young function and let $\bar{\Phi}$ be its dual. If $f \in L^\Phi$ and $g \in L^{\bar{\Phi}}$, then $fg \in L^1$, and*

$$\int |fg| \, d\mu \le 2\|f\|_\Phi \|g\|_{\bar{\Phi}}.$$

Proof of Theorem 10.2. It's enough to prove the theorem when $\|f\|_\Phi = \|g\|_{\bar{\Phi}} = 1$. By the construction of $\bar{\Phi}$, we have, for every x in our measure space,

$$|f(x)g(x)| \le \Phi(|f(x)|) + \bar{\Phi}(|g(x)|).$$

Integrating both sides with respect to $d\mu(x)$ gives

$$\int |f(x)g(x)| \, d\mu(x) \le 2,$$

and that's what we wanted.

The reader may have noted that, in several of the previous examples, we did not bother to find a precise formula for a given Young function, but were content with a "large x" estimate. The reason for this is that the only Orlicz spaces of interest to us will be those in which the underlying measure spaces are finite. For such spaces, the question of whether f belongs to L^Φ only depends on how $\Phi(x)$ grows as $x \to \infty$.

These finite measure spaces will, in fact, be very simple probability spaces.

Definition 10.2. *Let Φ be a Young function. If $Q \subset \mathbf{R}^d$ is a cube and $f : \mathbf{R}^d \mapsto \mathbf{R}$ is Lebesgue measurable, we define $\|f\|_{\Phi,Q}$ to be the the L^Φ norm of f where the underlying measure space is $|Q|^{-1}\chi_Q \, dx$. I.e.,*

$$\|f\|_{\Phi,Q} = \inf\{\lambda > 0 : \frac{1}{|Q|} \int_Q \Phi(|f|/\lambda) \, dx \le 1\}.$$

A few remarks are in order.
1. If $\Phi(x) = x^p$ $(1 \le p < \infty)$ then

$$\|f\|_{\Phi,Q} = \left(\frac{1}{|Q|} \int_Q |f|^p \, dx\right)^{1/p},$$

as the reader can (and should) verify. This $\|f\|_{\Phi,Q}$ gives a useful, locally averaged measure of the size of f.

2. If

$$\Phi(x) = \begin{cases} 0 & \text{if } 0 \le x \le 1; \\ \infty & \text{otherwise,} \end{cases}$$

then $\|f\|_{\Phi,Q}$ is the essential supremum of $f\chi_Q$. This $\|f\|_{\Phi,Q}$ does NOT give a useful, locally averaged measure of the size of f.

3. Just to get used to the notation, we'll look at what might or might not be an old friend. (If it's not, just take what I say on faith.) A function f is said to be of *bounded mean oscillation* ($f \in BMO$) if

$$\sup_{Q \subset \mathbf{R}^d} \frac{1}{|Q|} \int_Q |f - f_Q| \, dx < \infty,$$

and the supremum is denoted by $\|f\|_*$. The celebrated John-Nirenberg theorem states that, if $f \in BMO$, there is positive constant c_f such that

$$\frac{1}{|Q|} \int_Q \exp(c_f |f - f_Q|) \, dx \le 2$$

for all cubes Q (see [25] for a nice proof). Moreover, this c_f satisfies $c_f > A(d)/\|f\|_*$, where $A(d)$ is positive and only depends on the dimension d.

Let's restate this result in terms of local Orlicz norms. Define $\Phi_1(x) = x$ and $\Phi_2(x) = \exp(x) - 1$. The John-Nirenberg theorem becomes:

$$\sup_{Q \subset \mathbf{R}^d} \|f - f_Q\|_{\Phi_2,Q} \le C \sup_{Q \subset \mathbf{R}^d} \|f - f_Q\|_{\Phi_1,Q},$$

for a positive constant C that depends only on d.

In our opinion, this phrasing of the theorem is almost completely unintelligible; but, precisely because of that, unwinding it might be an illuminating exercise.

We will encounter local Orlicz norms in the context of *Orlicz maximal functions*.

Definition 10.3. *If Φ is a Young function and $f : \mathbf{R}^d \mapsto \mathbf{R}$ is Lebesgue measurable, the Orlicz maximal function of f, $M_\Phi(f)(x)$, is defined by:*

$$M_\Phi(f)(x) \equiv \sup_{Q:x \in Q} \|f\|_{\Phi,Q}.$$

If $\Phi(x) = x^r$ $(1 \le r < \infty)$ then

$$M_\Phi(f)(x) = \sup_{Q:x \in Q} \left(\frac{1}{|Q|} \int_Q |f|^r \, dt \right)^{1/r}.$$

This maximal function, which is obviously similar to $M_{r,d}$, is often denoted as $M_r(f)$, and we will follow that convention. Note that setting $\Phi = x$ $(r = 1)$ gives the usual Hardy-Littlewood maximal function.

By virtue of the generalized Hölder inequality (Theorem 10.2), we have the following interesting and useful result.

Theorem 10.3. *Let Φ be a Young function and let $\bar{\Phi}$ be its dual. Then the pointwise inequality*

$$M(fg)(x) \leq 2(M_\Phi(f)(x))(M_{\bar{\Phi}}(g)(x))$$

holds for all $x \in \mathbf{R}^d$, for all Lebesgue measurable f and g.

The reader knows that M_1 is a bounded sublinear operator on L^p for $p > 1$, but not for $p = 1$. In a similar fashion, M_r is bounded on L^p for $p > r$, but not for $p = r$. Now, mathematicians love to motivate theorems by calling questions "natural," and it is certainly natural to ask the following: *Given a Young function Φ, when does M_Φ define a bounded operator on L^p?* However, we can do better here. Our work on the square function and singular integral operators has yielded a collection of inequalities of the form

$$\int |T(f)|^p \, v \, dx \leq C \int |f|^p \, M^*(v) \, dx,$$

valid for all weights v and all $1 < p \leq 2$, where $M^*(\cdot)$ denotes some maximal operator depending on T and p. We left open the question of how to generalize such inequalities to $p > 2$. The following theorem plays an essential part in filling this gap.

Theorem 10.4. *Let Φ be a Young function and let $1 < p < \infty$. Then M_Φ is bounded on $L^p(\mathbf{R}^d)$ if*

$$\int_1^\infty \Phi(x) \frac{dx}{x^{p+1}} < \infty. \tag{10.1}$$

Remark. This theorem is due to Carlos Pérez [47], and we shall refer to 10.1 as the Pérez condition. The Pérez condition is also necessary for L^p boundedness. However, since all of our interest in the theorem concerns its sufficiency, we will hold off proving necessity until the end of the chapter.

Remark. The '1' at the lower limit of the integral 10.1 can be replaced by any positive number c.

Before proving the theorem, let's see what it says about some of the Young functions we've looked at.

1. If $\Phi(x) = x^r$, then Φ satisfies the Pérez condition 10.1 if and only if $r < p$, which is a classical result.

2. If $\Phi(x) \sim x^r (\log(1+x))^{sr}$ for large x, then the Theorem 10.4 says that M_Φ is bounded when $r < p$ and unbounded when $r > p$. What about when $r = p$? In that case,

$$\int_c^\infty \Phi(x) \frac{dx}{x^{p+1}} = \int_c^\infty \frac{1}{x(\log(1+x))^{-sp}} \, dx,$$

which is finite when $sp < -1$ and infinite otherwise. In other words, when $r = p$, a sufficiently large *negative* power of the logarithm can "mollify" x^r enough to make M_Φ bounded on L^p.

3. The Young function $\Phi(x) = \exp(x) - 1$ doesn't satisfy 10.1. It isn't hard to see that M_Φ isn't bounded on any L^p ($1 < p < \infty$). For $n = 0, 1, 2, 3, \ldots$, let $E_n \subset [n, n+1]$ be measurable. (The measures of the E_n's will be chosen presently.) Define $f = \sum \chi_{E_n}$. Then $\int |f|^p \, dx = \sum |E_n|$. For each n (the reader should check this),

$$\|f\|_{\Phi,[n,n+1]} = \frac{1}{\log(1 + 1/|E_n|)},$$

and therefore

$$M_\Phi f(x) \geq \frac{1}{\log(1 + 1/|E_n|)}$$

for all $x \in [n, n+1]$, implying that

$$\int (M_\Phi f)^p \, dx \geq \sum \left(\frac{1}{\log(1 + 1/|E_n|)} \right)^p.$$

Setting $|E_n| = 1/(n+1)^2$ gives an $f \in L^p$ for which $M_\Phi(f) \notin L^p$. (Question for the reader: Is this M_Φ bounded on L^∞?)

Proof of Theorem 10.4. We will first prove sufficiency of 10.1 for boundedness of the dyadic form of M_Φ; i.e., for

$$M_{\Phi,d}(f)(x) \equiv \sup_{Q:x\in Q\in\mathcal{D}_d} \|f\|_{\Phi,Q},$$

where the supremum is only taken over dyadic cubes containing x. The general case will follow from a slight modification of this argument.

Boundedness of $M_{\Phi,d}$ will follow from the inequality:

$$|\{x : M_{\Phi,d}(f)(x) > t\}| \leq C \int_{\{x:\ |f(x)|\geq ct\}} \Phi(|f(x)|/t) \, dx, \qquad (10.2)$$

valid for all $t > 0$, and for some positive constants C and c. That's because, once we have 10.2, we can multiply both sides by pt^{p-1}, integrate from 0 to infinity, and get:

$$\int (M_{\Phi,d}(f))^p \, dx = p \int_0^\infty t^{p-1} |\{x : M_{\Phi,d}(f)(x) > t\}| \, dt$$

$$\leq C \int_0^\infty t^{p-1} \left(\int_{\{x:\ |f(x)|\geq ct\}} \Phi(|f(x)|/t) \, dx \right) dt$$

$$= C \int |f(x)|^p \left(\int_c^\infty \frac{\Phi(u)}{u^{p+1}} \, du \right) dx$$

$$= C \int |f(x)|^p \, dx,$$

where the next-to-last line follows from Fubini-Tonelli and a substitution $u = |f|/t$.

The proof of 10.2 is not hard. Let N be a large number, and define

$$M_{\Phi,d,N}(f)(x) \equiv \sup_{\substack{Q:x\in Q\in\mathcal{D}_d \\ \ell(Q)\leq N}} \|f\|_{\Phi,Q}.$$

It will be enough to prove 10.2 with $M_{\Phi,d,N}(f)$ in place of $M_{\Phi,d}(f)$, so long as we can do so with constants C and c that do not depend on N. Let Q be any dyadic cube with $\ell(Q) \leq N$ and such that $\|f\|_{\Phi,Q} > t$. Now recall the definition of $\|f\|_{\Phi,Q}$ for a moment. Having $\|f\|_{\Phi,Q} > t$ implies that if $0 < \lambda \leq t$, then we must have

$$\frac{1}{|Q|} \int_Q \Phi(|f|/\lambda)\,dx > 1.$$

In particular,

$$\frac{1}{|Q|} \int_Q \Phi(|f|/t)\,dx > 1.$$

These dyadic cubes Q have *bounded diameters*. Therefore, each one is contained in a maximal dyadic Q_k satisfying $\|f\|_{\Phi,Q_k} > t$, and these cubes Q_k are disjoint. For each Q_k, we have

$$|Q_k| \leq \int_{Q_k} \Phi(|f|/t)\,dx. \tag{10.3}$$

Now, we would like to say, "Sum over k and we're done." Unfortunately, doing that would only yield

$$|\{x :\; M_{\Phi,d,N}f(x) > t\}| \leq \int \Phi(|f(x)|/t)\,dx,$$

which is not quite what we need.

Let $c > 0$ be such that $\Phi(c) < .5$; such a c exists because Φ is continuous at 0. Then

$$\int_{\{x\in Q_k:\; |f(x)|\leq ct\}} \Phi(|f|/t)\,dx \leq .5|Q_k|.$$

Combining this with 10.3 gives

$$\int_{\{x\in Q_k:\; |f(x)|\geq ct\}} \Phi(|f|/t)\,dx \geq .5|Q_k|,$$

and therefore

$$|Q_k| \leq 2 \int_{\{x\in Q_k:\; |f(x)|\geq ct\}} \Phi(|f|/t)\,dx.$$

If we now sum this inequality over k, we get 10.2 for $M_{\Phi,d,N}(f)$, which implies the L^p boundedness of $M_{\Phi,d}$.

We shall now prove sufficiency for the *non-dyadic* maximal function M_Φ. Our proof will follow (pretty closely) the one we gave for $M_{\Phi,d}$, and in fact will depend on it. We will show:

$$|\{x: \ M_\Phi(f)(x) > t\}| \le C|\{x: \ M_{\tilde{\Phi},d}(f)(x) > t\}|,$$

where $\tilde{\Phi} = A\Phi$ for some large constant A; and that will clearly suffice.

Suppose $M_\Phi(f)(x') > t > 0$. Then, repeating the argument given above, x' lies in a cube Q such that

$$\int_Q \Phi(|f|/t)\, dx \ge |Q|.$$

We will now use a simple geometrical fact: *For every cube $Q \subset \mathbf{R}^d$, there exist 3^d congruent, dyadic cubes $\{Q_i\}_1^{3^d}$, all satisfying $(1/2)\ell(Q) < \ell(Q_i) \le \ell(Q)$, and such that*

$$Q \subset \cup Q_i.$$

(It's easiest to see this when $d = 1$; the general case follows by covering Q 'one dimension at a time.' By the way, this was a suggested exercise in chapter 1.) At least one of these Q_i's must satisfy

$$\int_{Q_i} \Phi(|f|/t)\, dx \ge (1/3^d)|Q| \ge (1/3^d)|Q_i|,$$

or

$$\int_{Q_i} \tilde{\Phi}(|f|/t)\, dx \ge |Q_i|,$$

where $\tilde{\Phi} = 3^d\Phi$. This implies that $\|f\|_{\tilde{\Phi},Q_i} \ge t$, and thus that $M_{\tilde{\Phi},d}(f) \ge t$ on all of Q_i. But Q_i is not much smaller than Q, and it is close to Q; in fact, Q is contained in $20Q_i$, the twenty-fold dilate of Q_i. Therefore (we encourage the reader to fill in the details),

$$|\{x: \ M_\Phi(f)(x) > t\}| \le 20^d|\{x: \ M_{\tilde{\Phi},d}(f)(x) > t\}|,$$

finishing the proof of sufficiency.

We will be most interested in Orlicz maximal functions M_Φ such that $\Phi(x) \sim x^r(\log(1 + x))^{sr}$ for large x, with $r \ge 1$. Our most important applications of these will require a generalization of the Orlicz Hölder inequality proved above.

Suppose that A and B are Young functions with the property that, for some positive y, the function $B(xy) - A(x)$ is *bounded* on $[0, \infty)$; an example of such a pair is $B(x) = x$ and $A(x) = x^r$, with $r > 1$. Note that, in this example, $B(xy) - A(x)$ isn't just bounded for some y, but goes to negative infinity, as $x \to \infty$, for *all* y. The same thing will happen in every case of interest to us.

Define

$$C(y) \equiv \sup\{B(xy) - A(x): \ x \ge 0\}.$$

By following virtually the same argument we gave earlier, we can show that C is also a Young function. (Our extra hypothesis on A and B ensures that $C(y)$ will be finite for some positive y.) Moreover, it satisfies

$$B(xy) \leq A(x) + C(y)$$

for all x and y. The argument in the proof of the generalized Hölder inequality goes through with easy modifications to imply that, for any f and g,

$$\|fg\|_B \leq 2\|f\|_A\|g\|_C. \tag{10.4}$$

We note in passing that the pointwise inequality,

$$M_B(fg)(x) \leq 2M_A(f)(x)M_C(g)(x),$$

follows immediately.

Although inequality 10.4 is beautiful and rigorous, in many applications it is not very useful, because it can be difficult to find a formula for C from the definition we have given. The trouble with our definition is that it's too general: it applies to arbitrary Young functions, but all the interesting applications involve "nice" Young functions.

What is "nice"?

Claim: Suppose that A, B, and C are three strictly increasing, everywhere continuous Young functions. Let A^{-1}, B^{-1}, and C^{-1} be their inverse functions. If, for all $t > 0$,

$$A^{-1}(t)C^{-1}(t) \leq B^{-1}(t), \tag{10.5}$$

then, for all positive x and y,

$$B(xy) \leq A(x) + C(y). \tag{10.6}$$

Conversely, if A, B, and C satisfy 10.6 for all x and y , then, for all $t > 0$,

$$A^{-1}(t)C^{-1}(t) \leq B^{-1}(2t). \tag{10.7}$$

Proof of claim: Let $A(x) = s$ and $C(y) = v$, and set $t = s + v$. Then, since $t \geq s$ and $t \geq v$, and A^{-1} and C^{-1} are both increasing,

$$xy = A^{-1}(s)C^{-1}(v) \leq A^{-1}(t)C^{-1}(t) \leq B^{-1}(t),$$

implying

$$B(xy) \leq t = s + v = A(x) + C(y),$$

which is 10.6. Conversely, if 10.6 holds and $t > 0$, let $A(x) = C(y) = t$. Then 10.6 implies that

$$B(xy) \leq A(x) + C(y) = 2t,$$

and therefore that

$$A^{-1}(t)C^{-1}(t) = xy \le B^{-1}(A(x) + C(y)) = B^{-1}(2t).$$

Conclusion: 10.6 and 10.5 are essentially equivalent, for appropriate Young functions. Also, it isn't hard to see that if A, B, and C satisfy 10.7 then they almost satisfy 10.6; namely, $B(xy) \le 2(A(x) + C(y))$—which is good enough to yield a generalized Hölder's inequality (up to a factor of 2).

Given this, it's natural to ask why we didn't use 10.5 or 10.7 to define the dual Young function C. The reason is that it isn't obvious that a C satisfying 10.5 or 10.7 will automatically be a Young function. In other words, the supremum process based on 10.6 gives us existence of an appropriate C. Once we have it (and if we know it has the right properties), we can use 10.7 to estimate C. What is important to keep in mind is that these two kinds of inequalities yield three Orlicz norms satisfying 10.4, and they allow us to control the Orlicz norm of a product by the product of two other Orlicz norms.

Now let's look at some examples.

1. If we put $B(x) = x^r$ and $A(x) = x^p$, with $1 \le r < p < \infty$, then freshman calculus shows that $C(y)$ is a constant times y^q, where $q = \frac{pr}{p-r}$, or $1/q = 1/r - 1/p$; and we obtain, up to a constant factor, the classical Young's Inequality.

2. Let $B(x) = x(\log(e + x))^s$ for large x and suppose $A(x) = x^r$, with $r > 1$. In this case it's much easier to use 10.5. We can also simplify things by remembering that we only need to estimate $C(y)$ for large y. We make a simple observation about numbers: *If p is any real number, the inverse function of $x(\log(e + x))^p$ is approximately $t(\log(e + t))^{-p}$, at least when x or t is large.* Therefore, $B^{-1}(t)$ is essentially $t(\log(e + t))^{-s}$. Now, the inverse function of A is trivially $t^{1/r}$. So, to get 10.5, we want

$$C^{-1}(t) \sim t^{1/r'}(\log(e + t))^{-s},$$

which will follow if

$$C(x) \sim x^{r'}((\log(e + x))^{sr'}.$$

3. Let B be as in the last example, but now let $A(x) \sim x^r(\log(e + x))^k$, where $r > 1$ and k is a real number. The inverse function of A is essentially

$$\frac{t^{1/r}}{(\log(e + t))^{k/r}},$$

implying that a good choice for $C^{-1}(t)$ is

$$\frac{t^{1/r'}}{(\log(e + t))^{s-k/r}},$$

which, after some juggling, yields

$$C(x) \sim x^{r'}(\log(e + x))^{(s-k/r)r'}.$$

4. Finally, for an extreme example, let's take $B(x) = \exp(x) - 1$ and $A(x) = \exp(x^r) - 1$, with $r > 1$. It isn't hard to see that the right C will have the form $C(x) \sim \exp(c\,x^{r'})$ for large x.

Theorem 10.4 answers the question of what Orlicz maximal functions M_Φ are bounded on L^p, and it naturally raises another question: Given an M_Φ, what growth condition on f ensures that $M_\Phi(f)$ is locally integrable? We can make this more precise: *Given a Young function Φ, does there exist another Young function Φ^* such that, for all f and all cubes Q,*

$$\int_Q M_\Phi(f\chi_Q)\,dx \le C\|f\chi_Q\|_{\Phi^*}, \tag{10.8}$$

where C is some absolute constant? Our interest in this question is practical, because very soon we will need to know what happens when we take (Hardy-Littlewood) maximal functions of (Orlicz) maximal functions, and 10.8 and its converse (see below) provide the key estimates. We will only consider these inequalities for the special case when Φ is everywhere continuous and strictly increasing. As our examples and discussions show, those are not very restrictive conditions.

One well-known instance of 10.8 is provided by the pair of functions $\Phi(x) = x$ and $\Phi^*(x) = x \log(e + x)$. A proof of 10.8 for these functions is given in [53]. That proof is what motivates the following definition.

Definition 10.4. *If B is a strictly increasing, everywhere continuous Young function, we define*

$$B^*(x) = x\left(1 + \left(\int_1^x \frac{B(u)}{u^2}\,du\right)_+\right).$$

It is not hard to see that B^* is also a strictly increasing, everywhere continuous Young function. It is also useful to observe that there is a positive constant C such that $B(x) \le CB^*(x)$ for all x; and therefore that, for any f,

$$\|f\|_B \le c\|f\|_{B^*}.$$

Our theorem is:

Theorem 10.5. *If B is a strictly increasing, everywhere continuous Young function, then there is a constant C, depending on B, such that, for all cubes Q and all f,*

$$\frac{1}{|Q|}\int_Q M_B(f\chi_Q)\,dx \le C\|f\|_{B^*,Q}. \tag{10.9}$$

Before proving Theorem 10.5, we should check it. Let $B(x) = x^r$, where $1 < r < \infty$. Then

$$B^*(x) = x\left(1 + \left(\int_1^x u^{r-2}\,du\right)_+\right) = \begin{cases} x & \text{if } 0 \le x \le 1; \\ x + \frac{x^r - x}{r-1} & \text{if } x > 1, \end{cases}$$

and thus is $\sim x^r$ when x is large. This amounts to saying

$$\frac{1}{|Q|} \int_Q M_r(f) \, dx \leq C_{r,d} \left(\frac{1}{|Q|} \int_Q |f|^r \, dx \right)^{1/r},$$

which looks a little weird, but is classical: we essentially proved it in chapter 3 (Lemma 3.2). If instead we put $B(x) = x$, then

$$B^*(x) = x \left(1 + \left(\int_1^x \frac{1}{u} \, du \right)_+ \right) = x(1 + \log_+ x) \sim x \log(e + x),$$

for large x, which gives the result from [53].

Proof of Theorem 10.5. We take f to be non-negative. We will prove the theorem for the *dyadic* form of M_B, and under the assumption that Q is a dyadic cube with volume equal to 1. We will also assume that f's support is contained inside Q and that $\|f\|_{B^*} = 1$. Because of B^*'s continuity, the last statement means the same as $\int_Q B^*(f) \, dx = 1$. The integral we are trying to control is less than or equal to

$$|Q| + \int_1^\infty |\{x \in Q : \ M_B(f)(x) > t\}| \, dt. \tag{10.10}$$

The first term—$|Q|$—is clearly no problem. Earlier arguments imply that the integrand in the t integral is less than or equal to

$$\int_{\{x \in Q: \ f(x) > ct\}} B(f/t) \, dx,$$

where c is a small positive constant depending on B. Therefore our t integral is bounded by

$$\int_1^\infty \left(\int_{\{x \in Q: \ f(x) > ct\}} B(f/t) \, dx \right) dt = \int_Q \left(\int_1^{f(x)/c} B(f(x)/t) \, dt \right)_+ dx.$$

After a change of variables, the inner integral becomes

$$f(x) \left(\int_c^f \frac{B(u)}{u^2} \, du \right)_+ \leq c' B^*(f(x)),$$

where the constant c' comes in because c might be pretty small (though it won't be zero). When we integrate this inequality in x, and combine it with our earlier estimate 10.10, we get 10.9.

Readers of [53] will recall that, when $B(x) = x$, Theorem 10.5 has a converse. Essentially the same argument as in [53] yields the analogous converse for general B.

Theorem 10.6. *Let B and B^* be as in Theorem 10.5. Then, for any cube Q,*

$$\|f\|_{B^*,Q} \le C \frac{1}{|Q|} \int_Q M_B(f\chi_Q)\, dx.$$

Proof of Theorem 10.6. Again we assume that Q is dyadic, with volume 1. We also assume that our non-negative f has support contained in Q and that $\int_Q B^*(f)\, dx = \|f\|_{B^*} = 1$; the reader should notice that the preceding equation implies that $\frac{1}{|Q|} \int_Q f\, dx \le 1$. We will show that $\int_Q M_B(f)\, dx$ is bounded below by an absolute constant.

We have two cases to consider. Write:

$$1 = \int_Q B^*(f)\, dx = \int_{\{x\in Q: f\le 10\}} B^*(f)\, dx + \int_{\{x\in Q: f>10\}} B^*(f)\, dx. \quad (10.11)$$

One of the integrals on the far right must be $\ge 1/2$. Suppose it's the first one. We split it into two pieces, and remember that $B^*(x) = x$ when $x \le 1$:

$$\int_{\{x\in Q: f\le 10\}} B^*(f)\, dx = \int_{\{x\in Q: f\le 1/4\}} B^*(f)\, dx + \int_{\{x\in Q: 1/4<f\le 10\}} B^*(f)\, dx$$

$$\le (1/4) + B^*(10)|\{x \in Q: \ 1/4 < f(x)\}|.$$

We are assuming that the integral on the left is $\ge 1/2$, which implies that

$$|\{x \in Q: \ 1/4 < f(x)\}| \ge (4B^*(10))^{-1} > 0.$$

This implies a lower bound on $\|f\|_{B,Q}$; because, for any $\lambda > 0$,

$$\int_Q B(f/\lambda)\, dx \ge \int_{\{x\in Q: f\ge 1/4\}} B(f/\lambda)\, dx \ge B(1/(4\lambda))|\{x \in Q: \ 1/4 < f(x)\}|,$$

which goes to infinity as $\lambda \to 0$. So, if we're going to have $\int_Q B(f/\lambda)\, dx \le 1$, λ has to be $\ge$ some $\lambda_0 > 0$. But then (remember that $|Q| = 1$):

$$\int_Q M_B(f)\, dx \ge \|f\|_{B,Q} \ge \lambda_0,$$

and we're done—in *this* case.

Now we assume that the second of the far-right integrals in 10.11 is $\ge 1/2$. On the set $\{x \in Q: \ f(x) > 10\}$,

$$B^*(f(x)) = f(x) + f(x) \int_1^{f(x)} \frac{B(u)}{u^2}\, du \le Cf(x) \int_1^{f(x)} \frac{B(u)}{u^2}\, du,$$

where C is some positive constant depending only on B. Therefore, our hypothesis on the integral in 10.11 implies

$$\int_Q f(x) \left(\int_1^{f(x)} \frac{B(u)}{u^2} \, du \right)_+ dx \geq \int_{\{x \in Q : f(x) > 10\}} f(x) \left(\int_1^{f(x)} \frac{B(u)}{u^2} \, du \right)_+ dx$$

$$\geq c,$$

where c is a positive constant. We will now use this inequality to get a lower bound on $\int_Q M_B(f) \, dx$. Here it will be important that B is continuous and *strictly* increasing.

Let $t > 1$. By suitably normalizing B, we may assume that $B(1)=1$. Then $B(f(x)/t) > 1$ precisely on the set where $f(x) > t$. For almost every $x \in Q$ for which $f(x) > t$, there is a maximal dyadic subcube $Q_k \subset Q$ such that $x \in Q_k$ and

$$\frac{1}{|Q_k|} \int_{Q_k} B(f(s)/t) \, ds > 1.$$

(Having $Q_k \subset Q$ follows from having $\frac{1}{|Q|} \int_Q f \, dx \leq 1$.) Because of Q_k's maximality,

$$2^d \geq \frac{1}{|Q_k|} \int_{Q_k} B(f(s)/t) \, ds.$$

Therefore,

$$|\{x : M_B f(x) > t\}| \geq \sum_k |Q_k|$$

$$\geq c \sum \int_{Q_k} B(f(s)/t) \, ds$$

$$\geq c \int_{\{s \in Q : f(s) > t\}} B(f(s)/t) \, ds, \qquad (10.12)$$

where 10.12 follows from the fact that $f(s) \leq t$ almost everywhere off of $\cup Q_k$. When we integrate both ends of 10.12 with respect to t, from 1 to infinity, the left end gives a lower bound for $\int_Q M_B(f) \, dx$, while the right end is a constant times

$$\int_1^\infty \left(\int_{\{x \in Q : f(x) > t\}} B(f(x)/t) \, dx \right) dt.$$

But the "familiar" change of variables shows that the last integral is equal to

$$\int_Q f(x) \left(\int_1^{f(x)} \frac{B(u)}{u^2} \, du \right)_+ dx,$$

which is bounded below by an absolute constant. The theorem is proved.

If we put together Theorem 10.5 and Theorem 10.6, and work a little, we get the following characterization of how the Hardy-Littlewood maximal operator affects an Orlicz maximal operator.

Theorem 10.7. *Let B be a strictly increasing, everywhere continuous Young function, and let B^* be as in the statement of Theorem 10.5. Then, for any $f \in L^1_{\mathrm{loc}}(\mathbf{R}^d)$,*

$$M(M_B(f)) \sim M_{B^*}(f) \tag{10.13}$$

pointwise, with comparability constants depending only on B and d.

Proof of Theorem 10.7. We will prove 10.13 for the *dyadic* versions of the operators involved (all three of them), leaving the extension to the general case as an exercise.

Let $Q \in \mathcal{D}_d$, and write $f = f_1 + f_2$. where $f_1 = f\chi_Q$. By Theorem 10.5 and Theorem 10.6,

$$\frac{1}{|Q|} \int_Q M_B(f_1)\, dx \sim \|f\|_{B^*,Q}.$$

This implies that our result will now follow if we can only show that

$$\frac{1}{|Q|} \int_Q M_B(f_2)\, dx \leq C M_{B^*}(f)(x_0)$$

holds for all $x_0 \in Q$, for some absolute constant C. But *that* is a consequence of the fact that $M_B(f_2)$ is constant across Q. In fact, $M_B(f_2)$ equals $\|f_2\|_{B,Q_2}$ for some dyadic Q_2 containing Q. This last quantity satisfies

$$\|f_2\|_{B,Q_2} \leq \|f\|_{B,Q_2} \leq C\|f\|_{B^*,Q_2},$$

which proves the theorem (in the *dyadic* case).

We've already seen two important special cases of Theorem 10.5 and its converse: $B(x) = x$, $B^*(x) \sim x\log(e+x)$; and $B(x) = x^r$, $B^*(x) \sim x^r$ (when x and r are bigger than 1). Now we'll see another example—or, rather, a whole scale of examples.

For $\alpha \geq 0$, set $B_\alpha(x) = x(\log(e+x))^\alpha$; *this notation will stand until the end of the chapter.* We claim (we have left the proof as an exercise) that $(B_\alpha)^*(x) \sim B_{\alpha+1}(x)$. This equivalence means that the operation of applying the Orlicz maximal operator M_{B_α} induces a progression in this particular family of Young functions, in which every step bumps up the power of the logarithm. In symbols: for every cube Q, and for every $\alpha \geq 0$,

$$\int_Q M_{B_\alpha}(f\chi_Q)(\xi)\, d\xi \sim \|f\chi_Q\|_{B_{\alpha+1}}. \tag{10.14}$$

Theorem 10.7 lets us understand this scale of Orlicz spaces in terms of iterated maximal functions. The theorem implies that, if $k \geq 1$ is an integer, then $M^k(f) \sim M_{B_{k-1}}(f)$. It also implies that

$$M^k(M_{B_\alpha}(f)) \sim M_{B_{\alpha+k}}(f).$$

If we define $M^0(f)$ to be $|f|$, then this last equivalence is even true for $k = 0$.

The scale of spaces defined by $\{B_\alpha\}_{\alpha \geq 0}$ gives the promised answer to a problem we raised in chapter 3, when we studied two-weight inequalities for the square function. Let's recall our basic, one-dimensional result: If $0 < p < \infty$ and $\alpha > p/2$, then the weighted inequality

$$\int |f|^p \, v \, dx \leq C \int (S(f))^p \, w \, dx,$$

(where $S(f)$ is as defined by 2.8) will hold for all $f = \sum_I \lambda_I h_{(I)}$, finite linear sums of Haar functions, for a pair of weights v and w, if

$$\int_I v(x) \, (\log(e + v(x)/v_I))^\alpha \, dx \leq \int_I w(x) \, dx$$

holds for all dyadic intervals I. We asked, "Given v, what is a reasonable w?" and we said that $w = M_d^{[\alpha]+1}(v)$ would do. But we also said that this majorant had two serious defects: 1) its estimation required applying the M_d operator many times; 2) it was too big when α was not an integer (because the order of the maximal operator jumped up).

The theory of Orlicz spaces *seems* to provide a more flexible family of maximal functions, one that might give us a better (as in "easier" and "smaller") choice of w's. But we immediately encounter a difficulty: our two-weight condition speaks in terms of integral expressions involving v; Orlicz norms, on the other hand, are defined indirectly, as infima. Now, it is evident that

$$\int_Q v(x) \, (\log(e + v(x)/v_Q))^\alpha \, dx$$

should be about the same size as $\|v\|_{B_\alpha, Q}$ for any cube Q. That is the content of the next theorem.

Theorem 10.8. *For any $\alpha \geq 0$, any weight v, and any cube $Q \subset \mathbf{R}^d$,*

$$\frac{1}{|Q|} \int_Q v(x) \, (\log(e + v(x)/v_Q))^\alpha \, dx \sim \|v\|_{B_\alpha, Q}, \qquad (10.15)$$

with (approximate) proportionality constants that depend only on α.

Remark. The analogue of Theorem 10.8 fails catastrophically for Young functions that grow very fast, even when they are finite and everywhere continuous. Take $\Phi(x) = e^x - 1$, and consider $f(x) = |\log x|$ on $[0, 1]$. The function f has average value 1, and a quick computation shows that $\|f\|_{\Phi, [0,1]} = 2$. But

$$\int_0^1 \Phi(f(x)) \, dx = \int_0^1 \left(\frac{1}{x} - 1 \right) dx = \infty.$$

Proof of Theorem 10.8. There is nothing to prove if $\alpha = 0$, so we take $\alpha > 0$. We also assume $|Q| = v_Q = 1$. We need to prove two inequalities. One of them is easy. Set $T = \int_Q v(x) \, (\log(e + v(x)))^\alpha \, dx$. We note two facts about T. The first fact is that T is finite if and only if $v \in L^{B_\alpha}(Q)$ (this is because the logarithm grows so slowly); indeed, T is finite if and only if $\int_Q B_\alpha(v/\lambda)\,dx$ is finite for all $\lambda > 0$. Therefore we can assume $T < \infty$, since otherwise there's nothing to prove. The second fact is that $T \geq 1$. Putting these two facts together, we get:

$$\int_Q B_\alpha(v(x)/T)\,dx = \frac{\int_Q v(x)\,(\log(e + v(x)/T))^\alpha\,dx}{\int_Q v(x)\,(\log(e + v(x)))^\alpha\,dx} \leq 1,$$

implying $\|v\|_{B_\alpha,Q} \leq T$.

Now we need to show that $\|v\|_{B_\alpha,Q} \geq c_\alpha T$, where $c_\alpha > 0$. Because of v's normalization, $\|v\|_{B_\alpha,Q} \geq 1$, and so we may assume that T is huge (of a precise hugeness yet to be determined). Now,

$$
\begin{aligned}
T \quad &= \int_{\{x \in Q:(\log(e+v(x)))^\alpha \leq T/2\}} v(x)\,(\log(e + v(x)))^\alpha\,dx \\
&\quad + \int_{\{x \in Q:(\log(e+v(x)))^\alpha > T/2\}} v(x)\,(\log(e + v(x)))^\alpha\,dx,
\end{aligned}
$$

while $\int_Q v\,dx = 1$. Therefore,

$$\int_{\{x \in Q:(\log(e+v(x)))^\alpha > T/2\}} v(x)\,(\log(e + v(x)))^\alpha\,dx \geq T/2.$$

Set $E = \{x \in Q : (\log(e + v(x)))^\alpha > T/2\}$. If T is very big (and we assume it is), then on E we have $v(x) \geq c_1 \exp(c_2 T^{1/\alpha})$, where c_1 and c_2 depend on α. This implies that, for $x \in E$, $v(x)/T$ will be bigger than or equal to a constant times $\sqrt{v(x)}$.

Set $\lambda_0 = \|v\|_{B_\alpha,Q}$. Because of B_α's continuity,

$$\int_Q B_\alpha(v/\lambda_0)\,dx = 1,$$

which implies

$$\int_Q v(x)\,(\log(e + v(x)/\lambda_0))^\alpha\,dx = \lambda_0.$$

We have already noted that $\lambda_0 \leq T$. Therefore:

$$
\begin{aligned}
\lambda_0 &= \int_Q v(x)\,(\log(e + v(x)/\lambda_0))^\alpha\,dx \\
&\geq \int_E v(x)\,(\log(e + v(x)/\lambda_0))^\alpha\,dx \\
&\geq \int_E v(x)\,(\log(e + v(x)/T))^\alpha\,dx \\
&\geq \int_E v(x)\,(\log(e + c\sqrt{v(x)}))^\alpha\,dx \\
&\geq c'_\alpha \int_E v(x)\,(\log(e + v(x)))^\alpha\,dx \\
&\geq c_\alpha T,
\end{aligned}
$$

where the last inequality follows from the fact that $\int_E v(x) (\log(e + v(x)))^\alpha \, dx \geq T/2$. Theorem 10.8 is proved.

Theorem 10.8, combined with 10.14, finally yields the family of majorants we promised in chapter 3:

Corollary 10.1. *Let $0 < p < \infty$ and let $\alpha > p/2$. Set $\beta = \max(\alpha - 1, 0)$. There is a constant $C = C(p, \alpha)$ such that, for all $f = \sum_I \lambda_I h_{(I)}$, finite linear sums of Haar functions, and all weights v,*

$$\int |f|^p \, v \, dx \leq C \int (S(f))^p \, M_{B_\beta}(v) \, dx.$$

We will (almost) conclude this chapter with a proof and discussion of a remarkable extrapolation theorem due to David Cruz-Uribe and Carlos Pérez [14]. This is not really a theorem in Littlewood-Paley theory, but one that applies to very general pairs of operators.

Theorem 10.9. *Let ϕ_1 and ϕ_2 be two measurable functions defined on $\mathbf{R}^d$. Suppose that, for some fixed $\alpha \geq 0$ and $0 < p_0 < \infty$, the following holds: There is a fixed constant C_1 so that, for all weights v,*

$$\int |\phi_1|^{p_0} \, v \, dx \leq C_1 \int |\phi_2|^{p_0} \, M_{B_\alpha}(v) \, dx. \tag{10.16}$$

Then, for all $p_0 < p < \infty$ and $\epsilon > 0$, there is a constant C_2, depending only on α, ϵ, d, p_0, and p, such that, for all weights v,

$$\int |\phi_1|^p \, v \, dx \leq C_2 C_1 \int |\phi_2|^p \, M_{B_{(p/p_0)(\alpha+1)+\epsilon-1}}(v) \, dx. \tag{10.17}$$

What does this give us? Set $\phi_1 = f$ and $\phi_2 = S(f)$, where f is a finite linear sum of Haar functions. Let $p_0 < 2 \leq p < \infty$. Inequality 10.16 holds for $p_0 < 2$ and $\alpha = 0$. Theorem 10.9 implies that

$$\int |f|^p \, v \, dx \leq C_1 C_2 \int (S(f))^p \, M_{B_{p/p_0+\epsilon-1}}(v) \, dx$$

holds. If we take p_0 very close to 2, then p/p_0 is just a little larger than $p/2$, and the inequality amounts to having

$$\int |f|^p \, v \, dx \leq \tilde{C} \int (S(f))^p \, M_{B_{\beta-1}}(v) \, dx,$$

valid for $\beta > p/2 \geq 1$. This is Corollary 10.1, which (looking back at chapter 3) we obtained in a way that depended on a series of subtle estimates for the square function. Theorem 10.9 shows that it only depends on one such (albeit rather subtle) estimate.

Here is another application. Put $\phi_1 = S(f)$ and $\phi_2 = f$. We know that

$$\int (S(f))^2\, v\, dx \leq C_p \int |f|^2\, M_{B_0}(v)\, dx.$$

(Actually, we know it for all $1 < p \leq 2$, but right now we want to take p as big as possible.) Theorem 10.9 lets us extend this L^2 result to L^p $(p > 2)$ with no additional work. We get:

$$\int (S(f))^p\, v\, dx \leq C_p \int |f|^p\, M_{B_{(p/2)-1+\epsilon}}(v)\, dx.$$

Notice that, again, the "order" of the maximal function is essentially $p/2$.

And here is yet another application. If $T_\mathcal{K}$ is any classical Calderón-Zygmund operator, as discussed in the preceding chapter, then the following inequality holds for all $f \in C_0^\infty(\mathbf{R}^d)$, all weights v, and all $1 < p < 2$:

$$\int |T_\mathcal{K}(f)|^p\, v\, dx \leq C \int |f|^p\, M^2(v)\, dx. \tag{10.18}$$

(Of course, we also have

$$\int |T_\mathcal{K}(f)|^2\, v\, dx \leq C \int |f|^2\, M^3(v)\, dx,$$

but we will not need that fact.) Theorem 10.9 gives the promised extensions to all $1 < p < \infty$. Simply take $1 < p_0 < 2 \leq p < \infty$, where we assume that p_0 is very close to 2. We have that $M^2(v) \sim M_{B_1}(v)$, and therefore

$$\int |T_\mathcal{K}(f)|^{p_0}\, v\, dx \leq C \int |f|^{p_0}\, M_{B_1}(v)\, dx.$$

But Theorem 10.9 implies that, for any $\epsilon > 0$,

$$\int |T_\mathcal{K}(f)|^p\, v\, dx \leq C \int |f|^p\, M_{B_{2(p/p_0)+\epsilon-1}}(v)\, dx. \tag{10.19}$$

If p is an integer, we can take ϵ very small and p_0 close enough to 2 to make $2(p/p_0) + \epsilon - 1 < p$, implying $M_{B_{2(p/p_0)+\epsilon-1}}(v) \leq CM^{[p+1]}(v)$. If p is not an integer, similar choices of ϵ and p_0 yield the same inequality. Therefore, for any $T_\mathcal{K}$, any $1 < p < \infty$, and any weight v,

$$\int |T_\mathcal{K}(f)|^p\, v\, dx \leq C \int |f|^p\, M^{[p+1]}(v)\, dx. \tag{10.20}$$

Inequalities 10.19 and 10.20 can both be seen as natural extensions of 10.18; which one is more illuminating is, perhaps, a matter of taste.

Proof of Theorem 10.9. Set $r = p/p_0 > 1$ and let r' be r's dual exponent, so that $\frac{1}{r} + \frac{1}{r'} = 1$. The function $B_\alpha(x)$ satisfies $B_\alpha^{-1}(t) \sim t/(\log(e + t))^\alpha$ when t is large. We seek two Young functions $A(x)$ and $C(x)$ such that $A^{-1}(t)C^{-1}(t) = B_\alpha^{-1}(t)$ and such that the maximal operator M_C is bounded on $L^{r'}(\mathbf{R}^d, dx)$; and, we would like C to be, in some sense, as "large" as possible. The reason for this quest will become apparent soon (we hope).

Because of Theorem 10.4, a good choice for $C(x)$ is one for which

$$C(x) \sim \frac{x^{r'}}{(\log(e + x))^{sr'}},$$

where $sr' > 1$. This is the same thing as saying $sr > r - 1$, so let's fix some $\epsilon > 0$, and put $sr = r - 1 + \epsilon$. For such a C, $C^{-1}(t) \sim t^{1/r'}(\log(e + t))^s$, implying

$$A^{-1}(t) \sim \frac{t^{1/r}}{(\log(e + t))^{s+\alpha}}$$

and

$$A(x) \sim x^r(\log(e + x))^{r(s+\alpha)}.$$

The generalized Hölder inequality implies $M_{B_\alpha}(fg) \leq cM_A(f)M_C(g)$ for any functions f and g. Let h be a function in $L^{r'}(v)$ such that $\int |h|^{r'} v\, dx = 1$ and

$$\int |\phi_1|^p v\, dx = \left(\int |\phi_1|^{p_0} h v\, dx \right)^{p/p_0}.$$

Writing $hv = (v^{1/r})(hv^{1/r'})$, we get

$$\int |\phi_1|^{p_0} h v\, dx \leq C_1 \int |\phi_1|^{p_0} M_{B_\alpha}(hv)\, dx$$

$$\leq cC_1 \int |\phi_1|^{p_0} M_A(v^{1/r}) M_C(hv^{1/r'})\, dx$$

$$\leq cC_1(I)(II),$$

where

$$(I) = \left(\int |\phi_1|^p (M_A(v^{1/r}))^r\, dx \right)^{1/r}$$

$$(II) = \left(\int (M_C(hv^{1/r'}))^{r'}\, dx \right)^{1/r'}.$$

Because of our choice of $C(x)$,

$$\int (M_C(hv^{1/r'}))^{r'}\, dx \leq K \int \left| hv^{1/r'} \right|^{r'} dx$$

$$= K \int |h|^{r'} v\, dx = K.$$

The theorem will be proved if we can show that

$$(M_A(v^{1/r}))^r \le cM_{B_{(p/p_0)(\alpha+1)+\epsilon-1}}(v),$$

where, recall, $r = p/p_0$ and $rs = r - 1 + \epsilon$. Fortunately, this just involves (very carefully) unwinding the definitions of both sides of the inequality.

I claim that, for any cube Q, and any $\lambda_0 > 0$,

$$\int_Q A(v^{1/r}/\lambda_0)\,dx \sim \int_Q B_{r(s+\alpha)}(v/\lambda_0^r)\,dx,$$

with approximate proportionality constants depending only on r, α, and s. To see that the claim is true, write out both sides. The left-hand integral is

$$\int_Q (v/\lambda_0^r)(\log(e + v^{1/r}/\lambda_0))^{r(s+\alpha)}\,dx.$$

The right-hand integral is

$$\int_Q (v/\lambda_0^r)(\log(e + v/\lambda_0^r))^{r(s+\alpha)}\,dx.$$

The only difference between them is in the arguments of the logarithms. We now prove the claim by observing that $\log(e + t) \sim \log(e + t^r)$ is valid for all positive t: the relevant "t" here is $t = v^{1/r}/\lambda_0$.

The claim implies that $(\|v\|_{A,Q})^r \sim c_{r,\alpha,s}\|v\|_{B_{r(s+\alpha)},Q}$, and therefore that

$$(M_A(v^{1/r}))^r \le cM_{B_{r(s+\alpha)}}(v).$$

But $r(s + \alpha) = rs + r\alpha = r\alpha + r - 1 + \epsilon = (p/p_0)(\alpha + 1) + \epsilon - 1$. That finishes the proof of Theorem 10.9.

Theorem 10.9 is very powerful. However, it is easy to believe that it says more than it actually does. Early on, we proved that, for all $0 < p < \infty$, if v and w are two weights in $L^1_{\text{loc}}(\mathbf{R}^d)$ such that

$$\int_Q v(x)\,(\log(e + v(x)/v_Q))^\alpha\,dx \le \int_Q w(x)\,dx \qquad (10.21)$$

for all cubes Q, for some $\alpha > p/2$, then

$$\int |f|^p\,v\,dx \le C \int (S(f))^p\,w\,dx$$

for all f which are finite linear sums of Haar functions. Theorem 10.9 might seem to make all of these theorems (one for each p) follow from a result for only one p; or that, at least, having the result for one p_0, we should be able to infer the corresponding results for all $p > p_0$. That is not quite correct. Theorem 10.9 says that, given an inequality of the form

$$\int |T_1 f|^{p_0}\, v\, dx \leq C \int |T_2 f|^{p_0}\, \tilde{M}(v)\, dx,$$

where $\tilde{M}$ is some Orlicz maximal operator, we can infer inequalities of the form

$$\int |T_1 f|^p\, v\, dx \leq C \int |T_2 f|^p\, M^*(v)\, dx$$

for all $p > p_0$, where M^* is some other Orlicz maximal operator. However, it is possible that, for a given v, many weights w satisfy 10.21, and none of them are Orlicz maximal functions of v. For example, if $v(x)$ is the absolute value of a polynomial function, then $v \in A_\infty$, and therefore $w = cv$ satisfies 10.21, where c is a "big enough" constant. But, unless v itself is constant, *any* Orlicz maximal function of such a v will be identically infinite.

Appendix: Necessity of the Pérez Condition

We will show that, if $\Phi : [0, \infty] \mapsto [0, \infty]$ is a Young function such that

$$\int_1^\infty \Phi(t)\, \frac{dt}{t^{p+1}} = \infty,$$

where $p > 1$, then $M_\Phi(\chi_{[0,1)}) \notin L^p(\mathbf{R})$. We are only considering the $d = 1$ situation because it is a little easier to see what is going on there; the reader should have no trouble generalizing the argument to $\mathbf{R}^d$.

We need to consider two cases: either $\Phi(x_0) = \infty$ for some finite x_0, or $\Phi(x) < \infty$ for all $x \in \mathbf{R}$. We leave it to the reader to show that, if the first case happens, $M_\Phi(\chi_{[0,1)})$ is bounded below by a positive constant, and we are done.

Therefore, we assume that Φ is everywhere finite. Since Φ is a Young function, this implies that there is an $a > 0$ such that Φ is continuous and strictly increasing on $[a, \infty)$. Define:

$$\tilde{\Phi}(x) = \begin{cases} x\left(\frac{\Phi(a+1)}{a+1} \right) & \text{if } 0 \leq x \leq a+1; \\ \Phi(x) & \text{if } x \geq a+1. \end{cases}$$

Then $\tilde{\Phi}$ is a Young function that is everywhere continuous and strictly increasing, and which equals Φ for large enough x. Therefore:

$$\int_1^\infty \tilde{\Phi}(x) \frac{dx}{x^{p+1}} = \infty \qquad (10.22)$$

and—very important—$M_{\tilde{\Phi}}(f) \sim M_\Phi(f)$, for any f.

We now *redefine* Φ to equal $\tilde{\Phi}$.

We leave the proof of the next lemma to the reader.

Lemma 10.1. *If 10.22 holds, then*

$$\sum_k 2^{-kp}\Phi(2^k) = \infty. \tag{10.23}$$

For $k \geq 0$, set $I^k \equiv [0, 2^k)$. We encourage the reader to show that the following equation is true:

$$\|\chi_{[0,1)}\|_{\Phi, I^k} = \left(\Phi^{-1}(2^k)\right)^{-1}, \tag{10.24}$$

where the superscript -1 on the Φ denotes the inverse function, and the second -1 is the plain old multiplicative inverse. Equation 10.24 has the consequence that, for all $x \in I^{k+1} \setminus I^k$ ($k \geq 0$), $M_\Phi(\chi_{[0,1)})(x) \geq \left(\Phi^{-1}(2^{k+1})\right)^{-1}$, and therefore

$$\int (M_\Phi(\chi_{[0,1)})(x))^p \, dx \geq c \sum_1^\infty \frac{2^k}{(\Phi^{-1}(2^k))^p}. \tag{10.25}$$

We will prove necessity of 10.1 by showing that 10.23 forces the right-hand side of 10.25 to be infinite.

For each $j \geq 1$, let E_j be the set of $k \geq 1$ such that

$$2^j \leq \Phi^{-1}(2^k) < 2^{j+1}.$$

The right-hand side of 10.25 is bounded below by a constant times

$$\sum_{j \geq 1} 2^{-jp} \sum_{k: k \in E_j} 2^k.$$

I claim that, for sufficiently large j, each set E_j is non-empty. That will prove necessity, because then we will have, for some large N,

$$\sum_{j \geq 1} 2^{-jp} \sum_{k: k \in E_j} 2^k \geq \sum_{j \geq N} 2^{-jp}\Phi(2^j) = \infty.$$

To see that E_j is eventually non-empty, we note that its defining condition is equivalent to having

$$\Phi(2^j) \leq 2^k < \Phi(2^{j+1}).$$

The convexity of Φ implies that, for all j,

$$\frac{\Phi(2^j)}{2^j} \leq \frac{\Phi(2^{j+1})}{2^{j+1}},$$

hence

$$\Phi(2^{j+1}) \geq 2\Phi(2^j).$$

Let j be so large that the least k for which $\Phi(2^j) \leq 2^k$ satisfies $k \geq 1$. In other words, let j be so large that $2^{k-1} < \Phi(2^j) \leq 2^k$ holds for some $k \geq 1$. Then

$\Phi(2^{j+1}) \geq 2\Phi(2^j) > 2^k$, implying that this k belongs to E_j. That finishes the proof.

Remark. The sets E_j are eventually non-empty because Φ's convexity forces it to grow fast, which means that its inverse function, while going to infinity, must grow fairly slowly, and so cannot jump over too many intervals of the form $[2^j, 2^{j+1})$.

Exercises

10.1. Show that, if Φ is any Young function, then there is a positive constant C, depending only on Φ, such that

$$\|f\|_\Phi \leq C\|f\|_\infty$$

holds in all measure spaces. Use this to show that the operator M_Φ is always bounded on $L^\infty(\mathbf{R}^d)$.

10.2. Show that if B is a strictly increasing, everywhere continuous Young function, and we set

$$B^*(x) = x \left(1 + \left(\int_1^x \frac{B(u)}{u^2} \, du \right)_+ \right),$$

then B^* is also a strictly increasing, everywhere continuous Young function. Also prove that there is a positive constant C such that $B(x) \leq CB^*(x)$ for all x.

10.3. Prove the non-dyadic version of Theorem 10.7. (Hint: Once again, our attention centers on what happens on a cube Q. The argument follows the pattern of the dyadic case, but with some small adaptations. Decompose $f = f_1 + f_2$, where $f_1 = f\chi_{\tilde{Q}}$, with $\tilde{Q}$ being Q's triple. The estimation of $M_B(f_1)$ goes much as before. However, the function $M_B(f_2)$ is likely not constant across Q.)

10.4. Suppose that A and B are Young functions with the property that, for some positive y, the function $B(xy) - A(x)$ is bounded on $[0, \infty)$. Define, for $y \geq 0$,

$$C(y) \equiv \sup\{B(xy) - A(x) : \ x \geq 0\}.$$

Show that C is a Young function.

10.5. Give a precise statement and proof of the following: "If p is any real number, the inverse function of $x(\log(e+x))^p$ is approximately $t(\log(e+t))^{-p}$, when x or t is large."

10.6. Show that, for all $\alpha \geq 0$,

$$(B_\alpha)^*(x) \sim B_{\alpha+1}(x),$$

with approximate proportionality constants depending only on α.

10.7. (Due to F. Nazarov.) Show that, for all $\alpha \geq 0$ and $\beta \geq 0$, and all measurable $f : \mathbf{R}^d \mapsto \mathbf{R}$,

$$M_{B_\alpha}(M_{B_\beta}(f)) \sim M_{B_{\alpha+\beta+1}}(f)$$

pointwise, with approximate proportionality constants depending only on α, β, and d. One can interpret this as implying that the operators M_{B_α} and M_{B_β} "almost commute."

Notes

Our exposition of beginning Orlicz space theory is based on [45]; see also [16] and [34]. The first generalized Hölder inequality (Theorem 10.2) was proved in [58]. Its generalization (inequality 10.4) is from [45]. Theorem 10.4 is from [47]. Theorem 10.9, which powerfully extends results from [12], [46], and [62], is from [14]. The many relationships between M_B and the $\| \cdot \|_{B^*}$ norm were pointed out to the author by F. Nazarov [44].

11

Goodbye to Good-λ

In this chapter we work on $\mathbf{R}^1$, and we only consider dyadic intervals.

We recall a result (Theorem 3.8) we proved in chapter 3:

Theorem 11.1. *Let $0 < p < \infty$ and let $\tau > p/2$. Let v and w be weights such that*

$$\int_I v(x) \left(\log(e + v(x)/v_I)\right)^\tau dx \leq \int_I w(x)\, dx$$

for all $I \in \mathcal{D}$. Then, for all finite sums $f = \sum_I \lambda_I h_{(I)}$,

$$\int (f^*(x))^p\, v\, dx \leq C \int (S(f)(x))^p\, w\, dx,$$

where the constant C only depends on p and τ.

We proved Theorem 3.8 by means of some good-λ inequality estimates and a telescoping series trick. Thanks to our work on Orlicz spaces, we are now able to prove the theorem without either of those devices.

The following proof, which I learned from Fedor Nazarov, is very short, but it does require a few preliminaries.

If $I \in \mathcal{D}$ and $\beta > 0$, we let $Exp(I, \beta)$ be the family of non-negative, measurable functions ϕ, supported on I, such that

$$\frac{1}{|I|} \int_I \exp(\phi(x)^\beta)\, dx \leq e + 1.$$

(Why $e+1$? See below.) The set $Exp(I, \beta)$ is "morally" the unit ball of a local Orlicz space. Recall that there is a Young function Φ such that $\Phi(x) = \exp(x^\beta)$ when x is large. Now fix Φ. The reader can show (and we leave it as an exercise) that there are positive constants C_1 and C_2, depending only Φ, such that, for all non-negative, measurable ϕ, supported on I,

$$\phi \in Exp(I, \beta) \Rightarrow \|\phi\|_{\Phi, I} \leq C_1$$

and

$$\|\phi\|_{\Phi,I} \leq C_2 \Rightarrow \phi \in Exp(I,\beta).$$

Given these implications, the next theorem is easy to prove.

Theorem 11.2. *Let v be non-negative and locally integrable, and let $\alpha > 0$. For all $I \in \mathcal{D}$,*

$$\int_I v(x) \left(\log(e + v(x)/v_I)\right)^\alpha dx \sim \sup \left\{ \int_I v(x)\,\phi(x)\,dx : \; \phi \in Exp(I,1/\alpha) \right\},$$
(11.1)

with comparability constants that only depend on α.

Proof of Theorem 11.2. The function $\phi(x) = (\log(e + v(x)/v_I))^\alpha$ belongs to $Exp(I,1/\alpha)$, which implies one inequality (and also explains why we picked $e + 1$). For the other direction (which is the one we actually need), set $B_\alpha(x) = x(\log(e+x))^\alpha$, and let $\Phi(x) = \overline{B_\alpha}(x)$, its dual Young function. We know that, for large x, $\Phi(x) \sim \exp(cx^{1/\alpha})$, where c is some positive constant. (See #7 in our list of dual Young function examples.) This implies that, if $\phi \in Exp(I,1/\alpha)$, then $\|\phi\|_{\Phi,I} \leq C_\alpha$. Therefore, for any $\phi \in Exp(I,1/\alpha)$, our Orlicz space Hölder inequality implies:

$$\frac{1}{|I|} \int_I v(x)\,\phi(x)\,dx \leq 2C_\alpha \|v\|_{B_\alpha,I}.$$

But, according to Theorem 10.8,

$$\|v\|_{B_\alpha,I} \leq \frac{C}{|I|} \int_I v(x) \left(\log(e + v(x)/v_I)\right)^\alpha dx,$$

which finishes the proof.

Corollary 11.1. *Let f be an integrable function supported in $I \in \mathcal{D}$ and suppose that $S(f) \in L^\infty$. If v is a weight and $0 < p < \infty$ then*

$$\int_I |f|^p v\,dx \leq C(p)\|S(f)\|_\infty^p \int_I v(x) \left(\log(e + v(x)/v_I)\right)^{p/2} dx.$$

Proof of Corollary 11.1. We make the usual reductions—$I = [0,1)$, $v_I = 1$—and we add another: $\int_0^1 f\,dx = 0$. (The reader should check that the last one is legitimate.) We also assume that $\|S(f)\|_\infty = 1$.

There is a $c > 0$, independent of f, such that $cf \in Exp(I,2)$. Thus, $|cf|^p \in Exp(I,2/p)$. The result follows now from Theorem 11.2.

We shall *almost* re-prove Theorem 3.8, leaving to an exercise how to fill a small gap.

Theorem 11.3. *Let $0 < p < \infty$, $\tau > p/2$, and let v and w be two weights such that*

$$\int_I v(x) \left(\log(e + v(x)/v_I)\right)^\tau dx \leq \int_I w(x) \, dx$$

for all $I \in \mathcal{D}$. Then

$$\int |f(x)|^p \, v \, dx \leq C(\tau, p) \int (S(f))^p \, w \, dx$$

for all finite linear sums $f = \sum_I \lambda_I h_{(I)}$.

Proof of Theorem 11.3. Set $p_0 = 2\tau > p$. By Corollary 11.1, if g is supported on $I \in \mathcal{D}$ and $S(g) \in L^\infty$, then

$$\int_I |g|^{p_0} \, v \, dx \leq C(p_0) \|S(g)\|_\infty^{p_0} \int_I v(x) \left(\log(e + v(x)/v_I)\right)^\tau dx$$

$$\leq C(p_0) \|S(g)\|_\infty^{p_0} \int_I w(x) \, dx.$$

We will use this fact shortly.

Let $f = \sum \lambda_I h_{(I)}$, a finite linear sum of Haar functions. For every integer k, set $E_k = \{x \in \mathbf{R} : S(f) > 2^k\}$, and define $D_k = \{I \in \mathcal{D} : I \subset E_k, I \not\subset E_{k+1}\}$. The sets D_k are disjoint (some of them might be empty) and $\cup_k D_k$ contains $\{I \in \mathcal{D} : \lambda_I \neq 0\}$ as a subset. This implies that $\{x : f(x) \neq 0\} \subset \cup_k E_k$. Put

$$f_k \equiv \sum_{I \in D_k} \lambda_I h_{(I)}.$$

The preceding arguments imply that $f = \sum_k f_k$. We notice that E_k is equal to a disjoint union of some maximal intervals; say $E_k = \cup_1^n I_k^j$. Each f_k's support is contained in the corresponding E_k, and the integral of f_k is zero over each of these maximal intervals I_k^j. We observe that $S(f_k) \leq C2^k$ everywhere.

The bound on $S(f_k)$ and the argument at the beginning of the proof imply, for each j and k,

$$\int_{I_k^j} |f_k|^{p_0} \, v \, dx \leq C(p_0) 2^{kp_0} \int_{I_k^j} w(x) \, dx.$$

Summing over j, we get:

$$\int |f_k|^{p_0} \, v \, dx \leq C(p_0) 2^{kp_0} \sum_1^n w(I_k^j)$$

$$\leq C(p_0) 2^{kp_0} w(E_k). \tag{11.2}$$

We are now ready to begin the proof proper. We divide it into two cases: i) $0 < p \leq 1$ and ii) $1 < p < \infty$.

Case i).

$$\int |f|^p \, v \, dx = \int \left| \sum_k f_k \right|^p v \, dx \leq \sum_k \int |f_k|^p \, v \, dx = \sum_k \int_{E_k} |f_k|^p \, v \, dx$$

$$\leq \sum_k v(E_k)^{1-p/p_0} \left[\int_{E_k} |f_k|^{p_0} \, v \, dx \right]^{p/p_0}$$

$$\leq C \sum_k v(E_k)^{1-p/p_0} \left[2^{kp_0} w(E_k) \right]^{p/p_0}$$

$$\leq C \sum_k v(E_k)^{1-p/p_0} 2^{pk} w(E_k)^{p/p_0} \leq C \sum_k 2^{pk} w(E_k)$$

$$= C \sum_k 2^{pk} w(\{ x \in \mathbf{R} : S(f) > 2^k \})$$

$$\leq C \int (S(f))^p \, w \, dx,$$

which finishes case i).

Case ii). Here we require a more sophisticated analysis. We begin with the observation that, when $x \in E_l \setminus E_{l+1}$,

$$f(x) = \sum_{k: \, k \leq l} f_k(x). \tag{11.3}$$

Why is 11.3 true? For any x,

$$f(x) = \sum_{I \in \mathcal{D}: \, x \in I} \lambda_I h_{(I)}.$$

But, if $x \notin E_{l+1}$, x cannot belong to any $I \in D_r$ for which $r > l$. Thus, if $x \in E_l \setminus E_{l+1}$,

$$f(x) = \sum_{I \in \mathcal{D}: \, x \in I} \lambda_I h_{(I)} = \sum_{\substack{I \in \mathcal{D}: \, x \in I, \\ I \notin D_r, \, r > l}} \lambda_I h_{(I)}$$

$$= \sum_{\substack{I \in \mathcal{D}: \, x \in I, \\ I \in D_r, \, r \leq l}} \lambda_I h_{(I)} = \sum_{k: \, k \leq l} f_k(x),$$

which is 11.3.

Let $\epsilon > 0$, to be specified presently. We recall that $\{ x : \, f(x) \neq 0 \} \subset \cup_l E_l$. Then:

$$\int |f(x)|^p \, v \, dx = \sum_l \int_{E_l \setminus E_{l+1}} |f(x)|^p \, v \, dx$$

$$= \sum_l \int_{E_l \setminus E_{l+1}} \left| \sum_{k: \, k \leq l} f_k(x) \right|^p v \, dx$$

$$\leq C_{\epsilon,p} \sum_l \int_{E_l \setminus E_{l+1}} \left[\sum_{k:\ k \leq l} 2^{\epsilon(l-k)} |f_k(x)|^p v\, dx \right]$$

$$\leq C_{\epsilon,p} \sum_{k,l:\ k \leq l} 2^{\epsilon(l-k)} \int_{E_l} |f_k(x)|^p v\, dx$$

$$\leq C_{\epsilon,p} \sum_{k,l:\ k \leq l} 2^{\epsilon(l-k)} v(E_l)^{1-p/p_0} \left[\int_{E_l} |f_k(x)|^{p_0} v\, dx \right]^{p/p_0}$$

$$\leq C_{\epsilon,p} \sum_{k,l:\ k \leq l} 2^{\epsilon(l-k)} v(E_l)^{1-p/p_0} \left[\int_{E_k} |f_k(x)|^{p_0} v\, dx \right]^{p/p_0},$$

$$(11.4)$$

where the (very small) change in the last line comes from the fact that $E_l \subset E_k$ if $k \leq l$. Arguing as in case i), for each k we have

$$\left[\int_{E_k} |f_k(x)|^{p_0} v\, dx \right]^{p/p_0} \leq C 2^{pk} w(E_k)^{p/p_0}.$$

Since it is trivial that $v \leq w$ almost everywhere, we also have $v(E_l) \leq w(E_l)$ for all l. Set $\delta = 1 - p/p_0 > 0$. For each k and l such that $k \leq l$, we can write:

$$v(E_l)^{1-p/p_0} \left[\int_{E_k} |f_k(x)|^{p_0} v\, dx \right]^{p/p_0} \leq C w(E_l)^\delta 2^{pk} w(E_k)^{1-\delta}.$$

When we substitute this into 11.4 we get

$$\int |f(x)|^p v\, dx \leq C \sum_{k,l:\ k \leq l} 2^{\epsilon(l-k)} w(E_l)^\delta 2^{pk} w(E_k)^{1-\delta}$$

$$= C \sum_{k,l:\ k \leq l} 2^{-(p\delta-\epsilon)(l-k)} \left[2^{pl} w(E_l) \right]^\delta \left[2^{pk} w(E_k) \right]^{1-\delta}$$

$$\leq C \sum_{k,l:\ k \leq l} 2^{-(p\delta-\epsilon)(l-k)} \left[\delta 2^{pl} w(E_l) + (1-\delta) 2^{pk} w(E_k) \right],$$

$$(11.5)$$

where the last line follows from the fact that $a^\delta b^{1-\delta} \leq \delta a + (1-\delta) b$ when a and b are non-negative.

Now we have to sum the two pieces of 11.5, where these two pieces are

$$C \sum_{k,l:\ k \leq l} 2^{-(p\delta-\epsilon)(l-k)} \left[\delta 2^{pl} w(E_l) \right] \tag{11.6}$$

and

$$C \sum_{k,l:\ k \leq l} 2^{-(p\delta-\epsilon)(l-k)} \left[(1-\delta) 2^{pk} w(E_k) \right]. \tag{11.7}$$

The first piece is bounded by a constant times

$$\sum_l 2^{pl} w(E_l) \left[\sum_{k:\, k \le l} 2^{-(p\delta-\epsilon)(l-k)} \right].$$

The inner sum,

$$\left[\sum_{k:\, k \le l} 2^{-(p\delta-\epsilon)(l-k)} \right],$$

will be bounded by a constant independent of l if $p\delta - \epsilon > 0$; i.e., if $\epsilon < p\delta$. Choose $\epsilon = p\delta/2$. Plugging this back in, 11.6 is seen to be less than or equal to a constant times

$$\sum_l 2^{pl} w(E_l) \le C \int (S(f))^p \, w \, dx,$$

which is what we wanted.

The second piece is bounded by a constant times

$$\sum_k 2^{pk} w(E_k) \left[\sum_{l:\, l \ge k} 2^{-(p\delta-\epsilon)(l-k)} \right].$$

Our choice of ϵ now ensures that the inner sum,

$$\sum_{l:\, l \ge k} 2^{-(p\delta-\epsilon)(l-k)},$$

is bounded by a constant independent of k. Plugging it back in, 11.7 is seen to be bounded by a constant times

$$\sum_k 2^{pk} w(E_k),$$

which is just right.

Exercises

11.1. The reader has probably noticed the small gap between Theorem 3.8 and Theorem 11.3: the first theorem gives a bound for f^*, the second gives one for f. This exercise outlines the argument needed to fill the gap.

a) Let f be a measurable function supported in $I \in \mathcal{D}$, and suppose that $\int_I f \, dx = 0$ and $S(f) \in L^\infty$. Show that, if v is a weight and $0 < p < \infty$, then

$$\int_I (f^*(x))^p \, v \, dx \le C(p) \|S(f)\|_\infty^p \int_I v(x) \, (\log(e + v(x)/v_I))^{p/2} \, dx.$$

(Hint: Assume $\|S(f)\|_\infty \le 1$, and show that there is positive constant c, independent of f and I, such that $cf^* \in Exp(I, 2)$.)

b) For the $p \le 1$ case we have

$$(f^*(x))^p \le \sum_k (f_k^*(x))^p,$$

and we can proceed almost exactly as in the proof of Theorem 11.3, using part a) at the right point.

c) We have to be more careful when $p > 1$. Show that $f^* \equiv 0$ outside $\cup_l E_l$, and therefore

$$\int (f^*(x))^p \, v \, dx = \sum_l \int_{E_l \setminus E_{l+1}} (f^*(x))^p \, v \, dx.$$

The trick now is to show that, when $x \in E_l \setminus E_{l+1}$,

$$f^*(x) = \left(\sum_{k:\, k \le l} f_k \right)^* (x).$$

From this point the rest of the proof is like that of Theorem 11.3, with another timely use of part a).

11.2. Generalize the proofs of Theorem 11.3 and the extension mentioned in problem 11.1 to functions defined on $\mathbf{R}^d$.

Notes

The proof given here comes from handwritten notes by F. Nazarov [44].

A Fourier Multiplier Theorem

We will give an application of much of the material from the preceding chapters by presenting a proof of the Hörmander-Mihlin Multiplier Theorem. We warn the reader that this chapter assumes some familiarity with the theory of Fourier multipliers, as treated in [53].

If $m : \mathbf{R}^d \mapsto \mathbf{C}$ is a bounded measurable function, the *Fourier multiplier operator* associated to m, denoted T_m, is defined implicitly by the formula

$$\widehat{T_m(f)}(\xi) = m(\xi)\hat{f}(\xi),$$

for functions f belonging to some suitable test class (such as $\mathcal{C}_0^\infty(\mathbf{R}^d)$). We restrict ourselves to bounded functions because we want to ensure that T_m maps $L^2(\mathbf{R}^d)$ into itself. By the Plancherel Theorem, it is easy to see that the operator norm of $T_m : L^2 \mapsto L^2$ is exactly equal to $\|m\|_\infty$.

The Hörmander-Mihlin Multiplier Theorem states that, if m is sufficiently smooth away from 0 (in a manner to be made precise), the Fourier multiplier operator T_m is bounded on $L^p(\mathbf{R}^d)$ for all $1 < p < \infty$. When we say that T_m is bounded, we mean that there is a positive constant $C(m,p)$ such that, for all $f \in \mathcal{C}_0^\infty(\mathbf{R}^d)$,

$$\|T_m(f)\|_p \leq C(m,p)\|f\|_p.$$

Such an *a priori* estimate allows us to define a unique extension of T_m—satisfying the same bound—to all of L^p.

What we will actually prove is that, if m is smooth enough, then, for all $\epsilon > 0$, there is a constant C_ϵ so that, for all $f \in \mathcal{C}_0^\infty(\mathbf{R}^d)$ and all bounded, compactly supported weights v,

$$\int |T_m(f)|^2 \, v \, dx \leq C_\epsilon \int |f|^2 \, M_{B_{3+\epsilon}}(v) \, dx, \tag{12.1}$$

where $M_{B_{3+\epsilon}}(v)$ is as we defined it in chapter 10, and can be taken to be:

$$M_{B_{3+\epsilon}}(v)(x) \equiv \sup_{Q:\, x \in Q} \frac{1}{|Q|} \int_Q v(t) \left(\log(e + v(t)/v_Q)\right)^{3+\epsilon} dt.$$

Since the operator $M_{B_{3+\epsilon}}$ is bounded on all of the L^p spaces such that $1 < p \le \infty$, 12.1 will imply that T_m extends to a bounded operator on L^p for all $2 \le p < \infty$. An argument from duality will then yield boundedness for $1 < p < 2$.

We will now define the sorts of multipliers we are interested in. We recall the standard convention that, if $\alpha = (\alpha_1, \ldots, \alpha_d)$ is a vector of non-negative integers, then the *differential monomial* D^α is the operator defined by

$$\left(\frac{\partial^{\alpha_1}}{\partial x_1^{\alpha_1}}\right)\left(\frac{\partial^{\alpha_2}}{\partial x_2^{\alpha_2}}\right)\cdots\left(\frac{\partial^{\alpha_d}}{\partial x_d^{\alpha_d}}\right),$$

and the order of D^α, defined to be $\alpha_1 + \cdots + \alpha_d$, is denoted by $|\alpha|$. If $|\alpha| = 0$ then D^α is just the identity operator.

Definition 12.1. *We will say that $m : \mathbf{R}^d \setminus \{0\} \mapsto \mathbf{C}$ is a* normalized Hörmander-Mihlin multiplier *if m is bounded, infinitely differentiable, and if, for all differential monomials D^α of order $\le d/2 + 1$ and all $\xi \in \mathbf{R}^d \setminus \{0\}$,*

$$|D^\alpha m(\xi)| \le |\xi|^{-|\alpha|}.$$

Theorem 12.1. *Let m be a normalized Hörmander-Mihlin multiplier. For every $\epsilon > 0$ there is a constant $C(\epsilon, d)$ such that, for all $f \in \mathcal{C}_0^\infty(\mathbf{R}^d)$ and all bounded, compactly supported weights v,*

$$\int |T_m(f)|^2 \, v \, dx \le C(\epsilon, d) \int |f|^2 \, M_{B_{3+\epsilon}}(v) \, dx.$$

The proof of the theorem will require a lemma and its corollary.

Lemma 12.1. *Let $\beta > d/2$ and suppose that $g \in L^2(\mathbf{R}^d)$. Set $h(x) \equiv g(x)(1 + |x|^2)^{-\beta/2}$. If f is a Schwartz function and v is any weight,*

$$\int |h * f(x)|^2 \, v \, dx \le C\|g\|_2^2 \int |f(x)|^2 \, M(v) \, dx,$$

where the constant C only depends on β and d.

Proof of Lemma 12.1. We assume that f and g are non-negative. For any x,

$$h * f(x) = \int \frac{g(x - t)}{(1 + |x - t|^2)^{\beta/2}} f(t) \, dt$$

$$\le \left(\int (g(x - t))^2 \, dt\right)^{1/2} \left(\int \frac{(f(t))^2}{(1 + |x - t|^2)^\beta} \, dt\right)^{1/2}$$

$$= \|g\|_2 \left(\int \frac{(f(t))^2}{(1 + |x - t|^2)^\beta} \, dt\right)^{1/2},$$

and therefore

$$\int |h*f(x)|^2 \, v \, dx \le \|g\|_2^2 \int (f(t))^2 \left(\int \frac{v(x)}{(1+|x-t|^2)^\beta} \, dx \right) dt.$$

But

$$\int \frac{v(x)}{(1+|x-t|^2)^\beta} \, dx \le C(\beta,d) M(v)(t).$$

(See exercise 2.7.) That proves the lemma.

We note that the preceding lemma can easily be generalized. Namely, the same conclusion will hold, with the same constant C, if, for the same g and β, and for some $y > 0$, $h(x) \equiv y^{-d}g(x/y)(1+|x/y|^2)^{-\beta/2}$. The reason is that the Cauchy-Schwarz inequality will now yield:

$$h*f(x) \le \left(\int y^{-d}(g((x-t)/y))^2 \, dt \right)^{1/2} \left(\int y^{-d} \frac{(f(t))^2}{(1+|(x-t)/y|^2)^\beta} \, dt \right)^{1/2}$$

$$= \|g\|_2 \left(\int y^{-d} \frac{(f(t))^2}{(1+|(x-t)/y|^2)^\beta} \, dt \right)^{1/2},$$

and therefore

$$\int |h*f(x)|^2 \, v \, dx \le \|g\|_2^2 \int (f(t))^2 \left(\int y^{-d} \frac{v(x)}{(1+|(x-t)/y|^2)^\beta} \, dx \right) dt.$$

But $y^{-d}(1+|(x-t)/y|^2)^{-\beta}$ is simply an L^1 dilate of $(1+|(x-t)|^2)^\beta$; and so, by exercise 2.7 again,

$$\int y^{-d} \frac{v(x)}{(1+|(x-t)/y|^2)^\beta} \, dx \le C(\beta,d) M(v)(t).$$

Proof of Theorem 12.1. Let $\phi : \mathbf{R}^d \mapsto \mathbf{R}$ be a real, radial Schwartz function, chosen so that its Fourier transform is always non-negative and has support contained in $\{\xi : 1 \le |\xi| \le 2\}$. Let $\psi \in \mathcal{C}_0^\infty(\mathbf{R}^d)$ be real, radial, satisfy $\int \psi \, dx = 0$, and be co-normalized with ϕ so that

$$\int_0^\infty \hat\phi(y\xi) \, \hat\psi(y\xi) \, \frac{dy}{y} \equiv 1$$

for all $\xi \ne 0$.

Now let Φ be a *third* real, radial Schwartz function, chosen so that $\int \Phi \, dx = 0$ and $\hat\Phi(\xi) \equiv 1$ when $1 \le |\xi| \le 2$. This second condition has the consequence that, for all $y > 0$ and all $x \in \mathbf{R}^d$

$$(T_m(\Phi_y)) * \phi_y(x) = \Phi_y * (T_m(\phi_y))(x).$$

Since $f \in L^2$, we also have

$$(T_m(f)) * \phi_y(x) = (T_m(f)) * \phi_y * \Phi_y(x) = f * \Phi_y * (T_m(\phi_y))(x)$$

for the same ranges of y and x. By our work on the Calderón reproducing formula, the preceding equation implies that

$$T_m(f)(x) = \int_{\mathbf{R}_+^{d+1}} (f * \Phi_y * (T_m(\phi_y))(t)) \, \psi_y(x - t) \, \frac{dt \, dy}{y},$$

where the convergence of the integral is in the L^2 sense, taken over any compact-measurable exhaustion of $\mathbf{R}_+^{d+1}$. This means that we can take an increasing sequence of finite families $\mathcal{F}_1 \subset \mathcal{F}_2 \subset \cdots \subset \mathcal{D}_d$ for which $\cup_j \mathcal{F}_j = \mathcal{D}_d$ and such that the sequence of functions defined by

$$g_j(x) \equiv \sum_{Q \in \mathcal{F}_j} \int_{T(Q)} (f * \Phi_y * (T_m(\phi_y))(t)) \, \psi_y(x - t) \, \frac{dt \, dy}{y}$$

converges to $T_m(f)$ almost everywhere as $j \to \infty$. By Fatou's Lemma it will be enough to show that

$$\int |g_j(x)|^2 \, v \, dx \leq C(\epsilon, d) \int |f(x)|^2 \, M_{B_{3+\epsilon}}(v) \, dx$$

for each j.

Our earlier work implies that

$$\int |g_j(x)|^2 \, v \, dx \leq C(\epsilon, d) \sum_{Q \in \mathcal{D}_d} \frac{|\lambda_Q|^2}{|Q|} \int_{\tilde{Q}} v(x) \, (\log(e + v(x)/v_{\tilde{Q}}))^{1+\epsilon} \, dx,$$

where

$$\lambda_Q = \left(\int_{T(Q)} |f * \Phi_y * (T_m(\phi_y))(t)|^2 \, \frac{dt \, dy}{y} \right)^{1/2}.$$

It is clear that

$$\frac{1}{|Q|} \int_{\tilde{Q}} v(x) \, (\log(e + v(x)/v_{\tilde{Q}}))^{1+\epsilon} \, dx \leq C M_{B_{1+\epsilon}}(v)(t)$$

if $t \in Q$, and therefore

$$\sum_{Q \in \mathcal{D}_d} \frac{|\lambda_Q|^2}{|Q|} \int_{\tilde{Q}} v(x) \, (\log(e + v(x)/v_{\tilde{Q}}))^{1+\epsilon} \, dx$$

$$\leq C \int_{\mathbf{R}_+^{d+1}} |f * \Phi_y * (T_m(\phi_y))(t)|^2 \, M_{B_{1+\epsilon}}(v)(t) \, \frac{dt \, dy}{y}.$$

Let us now closely examine the function $T_m(\phi_y)$. I claim that it is a boundedly positive multiple of an L^1 dilate of a function of the form $g(x)(1 + |x|^2)^{-\beta/2}$, where $\|g\|_2 \leq 1$. *This will prove the theorem*, because the corollary to Lemma 12.1 will yield, for every $y > 0$,

$$\int_{\mathbf{R}^d} |f * \Phi_y * (T_m(\phi_y))(t)|^2 \, M_{B_{1+\epsilon}}(v)(t) \, dt$$
$$\leq C \int_{\mathbf{R}^d} |f * \Phi_y(t)|^2 \, M(M_{B_{1+\epsilon}}(v))(t) \, dt \leq \int_{\mathbf{R}^d} |f * \Phi_y(t)|^2 \, M_{B_{2+\epsilon}}(v)(t) \, dt.$$

Integrating in y will then give us:

$$\int_{\mathbf{R}^{d+1}_+} |f * \Phi_y * (T_m(\phi_y))(t)|^2 \, M_{B_{1+\epsilon}}(v)(t) \, \frac{dt \, dy}{y}$$
$$\leq C \int_{\mathbf{R}^d} (G_1(f)(t))^2 \, M_{B_{2+\epsilon}}(v)(t) \, dt \leq C \int_{\mathbf{R}^d} |f(t)|^2 \, M(M_{B_{2+\epsilon}}(v))(t) \, dt$$
$$\leq C \int_{\mathbf{R}^d} |f(t)|^2 \, M_{B_{3+\epsilon}}(v)(t) \, dt.$$

In the preceding inequalities, the reader should note how the maximal function, acting on v, gets bumped up at every stage.

Now for the claim. First consider $y = 1$, and set $h(x) = T_m(\phi)(x)$. The Fourier transform of h, $\hat{h}(\xi)$, has support contained inside $\{\xi : 1 \leq |\xi| \leq 2\}$ and satisfies

$$|D^\alpha \hat{h}(\xi)| \leq C$$

for some absolute constant C, and for all multi-indices α such that $|\alpha| \leq d/2 + 1$. Set $\beta = [d/2] + 1$, and note that $\beta > d/2$. We have just seen that $D^\alpha \hat{h} \in L^2$ for all α such that $|\alpha| \leq \beta$. Therefore, by taking inverse Fourier transforms,

$$(1 + |x|^2)^{\beta/2} h(x) \in L^2(\mathbf{R}^d),$$

implying

$$h(x) = g(x)(1 + |x|^2)^{-\beta/2}$$

for some $g \in L^2$. That takes care of things when $y = 1$. For the general case, we consider the function with Fourier transform equal to $\hat{k}(\xi) \equiv m(\xi)\hat{\phi}(y\xi)$. The support of $\hat{k}$ is contained inside $\{\xi : 1/y \leq |\xi| \leq 2/y\}$ and satisfies

$$|D^\alpha \hat{k}(\xi)| \leq C y^{|\alpha|}$$

for all α such that $|\alpha| \leq \beta$. Set $\lambda(x) \equiv y^d k(xy)$. Its Fourier transform, $\hat{\lambda}(\xi)$, is equal to $\hat{k}(\xi/y)$. This Fourier transform is supported inside $\{\xi : 1 \leq |\xi| \leq 2\}$ and satisfies

$$|D^\alpha \hat{\lambda}(\xi)| \leq C$$

for all α such that $|\alpha| \leq \beta$. Therefore the function λ (up to multiplication by a bounded constant) satisfies the hypotheses of Lemma 12.1. Now the corollary to the lemma finishes the proof.

Theorem 12.1 implies that, for every normalized Hörmander-Mihlin multiplier m, the associated Fourier multiplier operator T_m is bounded on $L^p(\mathbf{R}^d)$

when $2 \leq p < \infty$, with a norm that only depends on d and p. Now suppose that $1 < p < 2$, and let p' be p's dual exponent. If f and g are two functions in $\mathcal{C}_0^\infty(\mathbf{R}^d)$, then

$$\int_{\mathbf{R}^d} (T_m(f)(x)) \, \overline{g(x)} \, dx$$

makes sense and is equal to

$$\int_{\mathbf{R}^d} m(\xi) f(\xi) \, \overline{\hat{g}(\xi)} \, d\xi = \int_{\mathbf{R}^d} f(x) \, \overline{(T_{\bar{m}}(g)(x))} \, dx.$$

But $\bar{m}$ is also a normalized Hörmander-Mihlin multiplier. Therefore

$$\left| \int_{\mathbf{R}^d} (T_m(f)(x)) \, \overline{g(x)} \, dx \right| \leq C \|f\|_p \|g\|_{p'}.$$

This holds for all $g \in \mathcal{C}_0^\infty(\mathbf{R}^d)$, and therefore

$$\|T_m(f)\|_p \leq C \|f\|_p$$

for all $f \in \mathcal{C}_0^\infty(\mathbf{R}^d)$. But that is exactly what we meant by boundedness of T_m.

Notes

The proof of the Hörmander-Mihlin Theorem ([29] [41]) given here owes much to its presentation in [16]. In [36] a version of the theorem is proved for A_p weights.

13

Vector-Valued Inequalities

Many of our weighted-norm results extend, with little extra work, to vector-valued functions. In this chapter we will restrict our discussion to functions $\mathbf{f} : \mathbf{R} \mapsto \ell^2(\mathbf{N})$. This is not the only possible setting for vector-valued inequalities—one could also consider functions mapping into ℓ^r for any $0 < r < \infty$ (see [18] and [24])—but it is the one that shows up most often in applications, and it is the "natural" one to look at when working with square functions.

When we say that $\mathbf{f}$ maps into ℓ^2, we mean that

$$\mathbf{f}(x) = (f_j)_1^\infty,$$

where each f_j is Lebesgue measurable and, for almost every $x \in \mathbf{R}$,

$$\|\mathbf{f}(x)\| \equiv \left(\sum_k |f_j(x)|^2 \right)^{1/2}$$

is finite. We will say that $\mathbf{f}$ belongs to (vector-valued) $L^p(\mathbf{R})$ if $\|\mathbf{f}(x)\|$ (as a function of x) belongs to (ordinary) $L^p(\mathbf{R})$, and the L^p norm of $\mathbf{f}$ is simply $\|\|\mathbf{f}\|\|_p$. However, to save eyestrain, we will usually denote the L^p norm of $\mathbf{f}$ by $\|\mathbf{f}\|_p$. Similarly, we will say that a Lebesgue measurable $\mathbf{f}$ is locally integrable if $\|\mathbf{f}\|$ is. It is useful to note that, if $\mathbf{f}$ is locally integrable, then, for every bounded measurable set E,

$$\int_E \mathbf{f} \, dt \equiv \left(\int_E f_j(t) \, dt \right)_1^\infty$$

exists as a vector in ℓ^2. It is also very useful to note that

$$\left\| \int_E \mathbf{f} \, dt \right\| \leq \int_E \|\mathbf{f}\| \, dt. \tag{13.1}$$

We suggest that the reader prove these two statements.

The *dyadic square function* of such an **f**, denoted **Sf**, is defined by

$$\mathbf{Sf}(x) \equiv \left(\sum_1^\infty (S(f_j)(x))^2 \right)^{1/2},$$

where $S(f_j)$ denotes the familiar, scalar-valued, dyadic square function. We similarly define the *intrinsic square function* of **f**, $\mathbf{G}_\beta(\mathbf{f})$, by

$$\mathbf{G}_\beta(\mathbf{f})(x) \equiv \left(\sum_1^\infty (G_\beta(f_j)(x))^2 \right)^{1/2}.$$

We will prove inequalities of the form

$$\int \|\mathbf{f}\|^p \, v \, dx \leq C \int (\mathbf{Sf})^p \, w \, dx$$

and

$$\int (\mathbf{Sf})^p \, v \, dx \leq C \int \|\mathbf{f}\|^p \, w \, dx$$

(and analogously for $\mathbf{G}_\beta(\mathbf{f})$) for pairs of weights v and w. We observe that, to prove such inequalities, it is sufficient to prove them for functions of the form $\mathbf{f} : \mathbf{R} \mapsto \mathbf{R}^M$, so long as the constants C we obtain are independent of M. *Henceforth we will assume that* $\mathbf{f}(x) = (f_j(x))_1^M$, *where M is unspecified but is assumed to be large.*

In order to state our theorems, it will be convenient to define a new test class.

Definition 13.1. *A vector-valued function* $\mathbf{f}(x) = (f_j(x))_1^M$ *is said to be of finite type if every f_j is a finite linear sum of Haar functions.*

The following theorems are direct analogues of results we proved earlier.

Theorem 13.1. *Let $0 < p < \infty$ and suppose that $\tau > p/2$. If v and w are weights such that, for all dyadic intervals I,*

$$\int_I v(x) \left(\log(e + v(x)/v_I) \right)^\tau dx \leq \int_I w(x) \, dx,$$

then, for all $\mathbf{f} : \mathbf{R} \mapsto \mathbf{R}^M$ *of finite type,*

$$\int \|\mathbf{f}\|^p \, v \, dx \leq C \int (\mathbf{Sf})^p \, w \, dx,$$

where C only depends on τ and p.

Theorem 13.2. *Let $1 < p \leq 2$. There is a constant C, depending only on p, so that, for all weights v and all* $\mathbf{f} : \mathbf{R} \mapsto \mathbf{R}^M$,

$$\int (\mathbf{Sf})^p \, v \, dx \leq C \int \|\mathbf{f}\|^p \, M_d(v) \, dx.$$

As with their scalar-valued predecessors, Theorem 13.1 and Theorem 13.2 will motivate (and quickly lead to) non-dyadic extensions.

Theorem 13.3. *Let* $0 < p < \infty$ *and* $0 < \beta \le 1$, *and suppose that* $\tau > p/2$. *If* v *and* w *are weights such that, for all intervals* I,

$$\int_I v(x) \left(\log(e + v(x)/v_I)\right)^\tau dx \le \int_I w(x)\, dx, \qquad (13.2)$$

then, for all $\mathbf{f} : \mathbf{R} \mapsto \mathbf{R}^M$ *such that each component* f_j *lies in* $\mathcal{C}_0^\infty(\mathbf{R})$,

$$\int \|\mathbf{f}\|^p\, v\, dx \le C \int (\mathbf{G}_\beta(\mathbf{f}))^p\, w\, dx, \qquad (13.3)$$

where C *only depends on* τ, β, *and* p. *In particular, when* $p < 2$, *13.3 holds when* $w = M(v)$.

Theorem 13.4. *Let* $1 < p \le 2$ *and* $0 < \beta \le 1$. *There is a constant* C, *depending only on* p *and* β, *so that, for all weights* v *and all* $\mathbf{f} : \mathbf{R} \mapsto \mathbf{R}^M$,

$$\int (\mathbf{G}_\beta(\mathbf{f}))^p\, v\, dx \le C \int \|\mathbf{f}\|^p\, M(v)\, dx.$$

As in the scalar-valued setting, the two preceding theorems yield corollaries for A_p weights.

Corollary 13.1. *Let* $1 < p < \infty$ *and* $w \in A_p$. *Suppose that* $0 < \beta \le 1$. *There is a constant* $C = C(p, \beta, w)$ *such that, for all* $\mathbf{f} : \mathbf{R} \mapsto \mathbf{R}^M$ *such that each component* f_j *lies in* $\mathcal{C}_0^\infty(\mathbf{R})$,

$$\int \|\mathbf{f}\|^p\, w\, dx \le C \int (\mathbf{G}_\beta(\mathbf{f}))^p\, w\, dx.$$

Corollary 13.2. *Let* $1 < p < \infty$ *and* $w \in A_p$. *Suppose that* $0 < \beta \le 1$. *There is a constant* $C = C(p, \beta, w)$ *such that, for all* $\mathbf{f} : \mathbf{R} \mapsto \mathbf{R}^M$,

$$\int (\mathbf{G}_\beta(\mathbf{f}))^p\, w\, dx \le C \int \|\mathbf{f}\|^p\, w\, dx.$$

The proof of Theorem 13.1 depends on the following lemma.

Lemma 13.1. *Let* $0 < \rho < 2$. *There are positive constants* c_ρ *and* C_ρ *such that, if* I *is any dyadic interval,* $\mathbf{f}$ *is of finite type, with support contained in* I, $\int \mathbf{f}\, dx = 0$, *and* $\|\mathbf{Sf}\|_\infty \le 1$, *then, for all* $\lambda > 0$,

$$|\{x \in I : \|\mathbf{f}(x)\| > \lambda\}| \le C_\rho |I| \exp(-c_\rho \lambda^\rho).$$

Remark. The meaning of Lemma 13.1 is that, for a vector-valued function **f**, boundedness of the vector-valued dyadic square function implies that **f** is *nearly* exponentially-square integrable.

Proof of Lemma 13.1. We apply Corollary 10.1, putting $p = 2$ and letting $\alpha > 1$ (to be specified presently). For every f_j we get

$$\int_I |f_j|^2 \, v \, dx \leq C \int_I (S(f_j))^2 \, M_{B_{\alpha-1}}(v) \, dx,$$

for all weights v supported on I, and therefore, after summing on j,

$$\int_I \|\mathbf{f}\|^2 \, v \, dx \leq C \int_I (\mathbf{S}\mathbf{f})^2 \, M_{B_{\alpha-1}}(v) \, dx \leq C \int_I M_{B_{\alpha-1}}(v) \, dx$$

$$\leq C \int_I v(x) \, (\log(e + v(x)/v_I))^\alpha \, dx,$$

where we have used Theorem 10.5 and Theorem 10.8 to infer the last inequality. Now set $E_\lambda \equiv \{x \in I : \|\mathbf{f}(x)\| > \lambda\}$ and put $v = \chi_{E_\lambda}$. An easy argument using Chebyshev's inequality yields

$$|E_\lambda| \leq C_\alpha |I| \exp(-c_\alpha \lambda^{2/\alpha}).$$

We conclude the proof by setting $\alpha = 2/\rho$.

The proof of the next corollary is very much like that of Corollary 11.1, and we omit it.

Corollary 13.3. *Let $0 < \rho < 2$ and $0 < p < \infty$. There is a positive constant C, depending only on p and ρ, such that, if I is any dyadic interval, $\mathbf{f}$ is of finite type, with support contained in I, $\int \mathbf{f} \, dx = 0$, and $\mathbf{S}\mathbf{f} \in L^\infty$, then, for weights v,*

$$\int_I \|\mathbf{f}\|^p \, v \, dx \leq C \|\mathbf{S}\mathbf{f}\|_\infty^p \int_I v(x) \, (\log(e + v(x)/v_I))^{p/\rho} \, dx.$$

Proof of Theorem 13.1. The proof follows the lines of the proof of Theorem 11.3, with only trivial modifications. We will sketch the outline and leave it to the reader to fill in the details.

Given $\tau > p/2$, pick $0 < \rho < 2$ such that $\tau > p/\rho$. Set $p_0 = \rho\tau > p$.

Much as in the proof of Theorem 11.3, for every integer k, we set $E_k = \{x \in \mathbf{R} : \mathbf{S}\mathbf{f} > 2^k\}$, and define $D_k = \{I \in \mathcal{D} : I \subset E_k, \ I \not\subset E_{k+1}\}$. We may write each E_k as a disjoint union $\cup_1^n I_k^l$ of maximal dyadic intervals. We are assuming that every f_j $(1 \leq j \leq M)$ has the form $f_j = \sum \lambda_{I,j} h_{(I)}$, which is a finite sum. For each j and k define

$$(f_j)_k \equiv \sum_{I \in D_k} \lambda_{I,j} h_{(I)},$$

and set $\mathbf{f}_k = ((f_j)_k)_1^M$. Then $\mathbf{f} = \sum_k \mathbf{f}_k$. (Note that this is actually a finite sum.) We define $\mathbf{Sf}_k$ to be the square function of $\mathbf{f}_k$; i.e.,

$$\mathbf{Sf}_k = \left(\sum_1^M ((S(f_j)_k))^2 \right)^{1/2}.$$

This decomposition of $\mathbf{f}$ is exactly analogous with the decomposition $f = \sum f_k$ we used in the proof of Theorem 11.3. The support of each $\mathbf{f}_k$ is contained in the corresponding E_k, and the integral of $\mathbf{f}_k$ (hence of each $(f_j)_k$) is zero over every interval I_k^l. We also observe that $\mathbf{Sf}_k \leq C2^k$ everywhere. Because of Corollary 13.3, for every I_k^l, we have

$$\int_{I_k^l} \|\mathbf{f}_k\|^{p_0} \, v \, dx \leq C2^{kp_0} \int_{I_k^l} v(x) \, (\log(e + v(x)/v_{I_k^l}))^{p_0/p} \, dx$$

$$= C2^{kp_0} \int_{I_k^l} v(x) \, (\log(e + v(x)/v_{I_k^l}))^\tau \, dx$$

$$= C2^{kp_0} w(I_k^l) \, dx,$$

where the last line comes from our hypothesis on v and w. When we sum over l, we get

$$\int \|\mathbf{f}_k\|^{p_0} \, v \, dx \leq C(p_0) 2^{kp_0} \sum_1^n w(I_k^l)$$

$$\leq C(p_0) 2^{kp_0} w(E_k).$$

But this is nothing but a vector-valued version of inequality 11.2. From this point the proof is practically identical to that of Theorem 11.3. We omit the details.

Proof of Theorem 13.2. The first step is to show that

$$\int (\mathbf{Sf})^2 \, v \, dx \leq C \int \|\mathbf{f}\|^2 \, M_d(v) \, dx,$$

which follows immediately from the scalar-valued result. The other step is to show the weak-type inequality,

$$v(\{x : \ \mathbf{Sf}(x) > \lambda\}) \leq \frac{C}{\lambda} \int \|\mathbf{f}\| \, M_d(v) \, dx.$$

The proof of this is almost a verbatim repetition of the proof of Theorem 3.10, and we will only sketch it. As in the scalar-valued case, we will assume that, for every $\epsilon > 0$, there is an $R > 0$ such that, if I is any interval satisfying $\ell(I) > R$, then

$$\frac{1}{|I|} \int_I \|\mathbf{f}\| \, dt < \epsilon.$$

(The treatment of the general case is almost exactly as with scalar-valued f's: see exercise 3.6.) Let Ω_λ be the set $\{x : \mathbf{Sf}(x) > \lambda\}$, and write $\Omega_\lambda = \cup I_j^\lambda$, a disjoint union of the maximal dyadic intervals such that

$$\frac{1}{|I_j^\lambda|} \int_{I_j^\lambda} \|\mathbf{f}\| \, dx > \lambda.$$

As in the proof of Theorem 3.10, we only need to show that

$$v(\{x \notin \Omega_\lambda : \mathbf{Sf}(x) > \lambda\}) \le \frac{C}{\lambda} \int \|\mathbf{f}\| \, M_d(v) \, dx.$$

Write $\mathbf{f} = \mathbf{g} + \mathbf{b}$, where

$$\mathbf{g}(x) = \begin{cases} \mathbf{f}(x) & \text{if } x \notin \cup_k I_k^\lambda; \\ \frac{1}{|I_k^\lambda|} \int_{I_k^\lambda} \mathbf{f}(t) \, dt & \text{if } x \in I_k^\lambda. \end{cases}$$

By virtue of inequality 13.1, the vector-valued function $\mathbf{g}$ satisfies $\|\mathbf{g}\|_\infty \le 2\lambda$, and, if $x \in I_j^\lambda$,

$$\|\mathbf{g}(x)\| \le \frac{1}{|I_j^\lambda|} \int_{I_j^\lambda} \|\mathbf{f}\| \, dt.$$

As with Theorem 3.10, $\mathbf{Sf} \le \mathbf{Sg} + \mathbf{Sb}$, and the support of $\mathbf{Sb}$ is entirely contained in Ω_λ. Therefore, it is enough to show

$$v(\{x \notin \Omega_\lambda : \mathbf{Sg}(x) > \lambda\}) \le \frac{C}{\lambda} \int \|f\| \, M_d(v) \, dx, \tag{13.4}$$

which is merely a vector-valued rephrasing of 3.28 from the proof of Theorem 3.10. The proof of 13.4 is almost identical to that of 3.28, and we leave it to the reader.

Proof of Theorem 13.3. Recall that we use $3\mathcal{D}$ to denote the family of concentric triples of the dyadic intervals $I \in \mathcal{D}$. Following our procedure from chapter 5 (Theorem 5.3), we can write $3\mathcal{D}$ as a disjoint union $\mathcal{G}_1 \cup \mathcal{G}_2 \cup \mathcal{G}_3$, where each $\mathcal{G}_i$ is a good family. We will fix our attention on $\mathcal{G}_1$, which we will simply refer to as $\mathcal{G}$. Let $\psi \in \mathcal{C}_0^\infty(\mathbf{R})$ be real, radial (i.e., even), supported in $[-1, 1]$, have integral equal to 0, and satisfy the usual normalization, to wit:

$$\int_0^\infty |\hat{\psi}(y\xi)|^2 \, \frac{dy}{y} \equiv 1$$

for all $\xi \ne 0$. If $g \in \mathcal{C}_0^\infty(\mathbf{R})$ and $\mathcal{F} \subset \mathcal{G}$ is any finite family, we set

$$g_\mathcal{F}(x) = \sum_{I: \tilde{I} \in \mathcal{F}} \int_{T(I)} (g * \psi_y(t)) \, \psi_y(x - t) \, \frac{dt \, dy}{y},$$

where we may always assume that the sum converges almost everywhere[1] and in L^2. If $\mathbf{f} : \mathbf{R} \mapsto \mathbf{R}^M$ is such that each of its components f_j lies in $\mathcal{C}_0^\infty(\mathbf{R})$, we define

$$\mathbf{f}_\mathcal{F} \equiv ((f_j)_\mathcal{F})_1^M.$$

Our earlier work on the convergence of the Calderón reproducing formula implies that the theorem will follow if we can show

$$\int \|\mathbf{f}_\mathcal{F}\|^p \, v \, dx \le C \int (\mathbf{G}_\beta(\mathbf{f}))^p \, w \, dx, \qquad (13.5)$$

so long as the constant C does not depend on $\mathcal{F}$.

Showing 13.5 does not take long. Each $(f_j)_\mathcal{F}$ is the L^2 and almost everywhere[2] limit of a sequence of finite sums $\sum_{I: \; \tilde{I} \in \mathcal{F}} \lambda_I(f_j) h_{\mathcal{F},(\tilde{I})}$, where the functions $h_{\mathcal{F},(\tilde{I})}$ are $\mathcal{F}$-adapted Haar functions as defined in chapter 4, and $\lambda_I(f_j) = \int f_j \, h_{\mathcal{F},(\tilde{I})} \, dx$. Still following the pattern of chapter 4, we can define

$$\mathbf{S}_\mathcal{F}(\mathbf{f}_\mathcal{F}) \equiv \left(\sum_1^M (S_\mathcal{F}((f_j)_\mathcal{F}))^2 \right)^{1/2},$$

which is really much easier to understand than it looks. Each $S_\mathcal{F}((f_j)_\mathcal{F})$ is merely the $\mathcal{F}$-adapted *dyadic* square function of $(f_j)_\mathcal{F}$. Our work in chapter 4 (Theorem 4.1) implies that, for each j,

$$S_\mathcal{F}((f_j)_\mathcal{F}) \le C \, \tilde{S}_{sd,\mathcal{F}}((f_j)_\mathcal{F})$$

pointwise, for an absolute constant C, where

$$\tilde{S}_{sd,\mathcal{F}}((f_j)_\mathcal{F})(x) = \left(\sum_{I: \; \tilde{I} \in \mathcal{F}} \frac{|\tilde{\lambda}_{(\tilde{I})}(f_j)|^2}{|\tilde{I}|} \chi_{\tilde{I}}(x) \right)^{1/2},$$

and

$$\tilde{\lambda}_{(\tilde{I})}(f_j) = \left(\int_{T(I)} |f_j * \psi_y(t)|^2 \frac{dt \, dy}{y} \right)^{1/2}.$$

At the beginning of chapter 6 we observed that $\tilde{S}_{sd,\mathcal{F}}((f_j)_\mathcal{F})$ is pointwise dominated by $S_{\psi,\alpha}(f_j)$ for sufficiently large $\alpha > 0$. Later in chapter 6, we also observed (and left as an exercise) that $S_{\psi,\alpha}(f_j) \le C G_\beta(f_j)$ pointwise, with a constant C depending on α, β, ψ, and d. Putting all of this together, we can now assert that

$$\mathbf{S}_\mathcal{F}(\mathbf{f}_\mathcal{F}) \le C \mathbf{G}_\beta(\mathbf{f}) \qquad (13.6)$$

---

[1] Actually, it converges uniformly for $g \in \mathcal{C}_0^\infty(\mathbf{R})$, but we don't need that.
[2] Uniform again!

pointwise. Since each $(f_j)_{\mathcal{F}}$ is the almost everywhere[3] limit of a sequence of finite sums $\sum_{I:\ \tilde{I}\in\mathcal{F}}\lambda_I(f_j)h_{\mathcal{F},(\tilde{I})}$, a repetition of the proof of Theorem 13.1, adapted now to the family $\mathcal{G}$, yields

$$\int\|\mathbf{f}_{\mathcal{F}}\|^p\,v\,dx\le C\int(\mathbf{S}_{\mathcal{F}}(\mathbf{f}_{\mathcal{F}}))^p\,w\,dx,$$

so long as 13.2 holds for all intervals. The pointwise estimate 13.6 lets us finish the proof.

Proof of Theorem 13.4. The L^2 inequality,

$$\int(\mathbf{G}_\beta(\mathbf{f}))^2\,v\,dx\le C\int\|\mathbf{f}\|^2\,M(v)\,dx,$$

follows immediately from the scalar-valued result. All we need to finish the proof is the weak-type bound,

$$v\left(\{x:\ \mathbf{G}_\beta(\mathbf{f})(x)>\lambda\}\right)\le\frac{C}{\lambda}\int\|\mathbf{f}\|\,M(v)\,dx.$$

We assume, as usual, that $\mathbf{f}$ has the property that, for every $\epsilon>0$, there is an $R>0$ such that, if I is any interval with $\ell(I)>R$, then

$$\frac{1}{|I|}\int_I\|\mathbf{f}\|\,dt<\epsilon,$$

and leave the general case as an exercise.

For $\lambda>0$, let $\{I_k^\lambda\}_k$ be the maximal dyadic intervals I such that

$$\frac{1}{|I|}\int_I\|\mathbf{f}\|\,dt>\lambda,$$

and set $\Omega=\cup_k\tilde{I}_k^\lambda$. Our work in chapter 6 (Step 3 in the proof of Theorem 6.1) implies that we only have to show

$$v\left(\{x\notin\Omega:\ \mathbf{G}_\beta(\mathbf{f})(x)>\lambda\}\right)\le\frac{C}{\lambda}\int\|\mathbf{f}\|\,M(v)\,dx.$$

Much as in the scalar case, define

$$\mathbf{g}(x)=\begin{cases}\mathbf{f}_{I_k^\lambda}&\text{if }x\in I_k^\lambda;\\\mathbf{f}(x)&\text{if }x\notin\cup_kI_k^\lambda,\end{cases}$$

and put $\mathbf{b}=\mathbf{f}-\mathbf{g}$. Our problem reduces to showing

$$v\left(\{x\notin\Omega:\ \mathbf{G}_\beta(\mathbf{g})(x)>\lambda/2\}\right)\le\frac{C}{\lambda}\int\|\mathbf{f}\|\,M(v)\,dx\qquad(13.7)$$

---

[3] See preceding footnotes.

and

$$v\left(\{x \notin \Omega : \ \mathbf{G}_\beta(\mathbf{b})(x) > \lambda/2\}\right) \le \frac{C}{\lambda} \int \|\mathbf{f}\| \, M(v) \, dx.$$

The function $\mathbf{g}$ satisfies the same bounds as the function $\mathbf{g}$ we encountered in the proof of Theorem 13.2. It is now straightforward (with the help of Lemma 6.1 from chapter 6) to show 13.7; we leave this as an exercise.

We may further decompose $\mathbf{b}$ into $\sum_k \mathbf{b}_k$, where $\mathbf{b}_k = \mathbf{b}\chi_{I_k^\lambda}$. These functions also satisfy bounds and equations similar to those of their scalar counterparts:

$$\int \mathbf{b}_k(t) \, dt = 0$$

$$\frac{1}{|I_k^\lambda|} \int \|\mathbf{b}_k(t)\| \, dt \le C \frac{1}{|I_k^\lambda|} \int_{I_k^\lambda} \|\mathbf{f}(t)\| \, dt$$

$$\le C\lambda.$$

The first formula means that every component of $\mathbf{b}_k$ has integral equal to 0.

Much as in the proof of Theorem 6.1 (see 6.9 and 6.10), our result will follow immediately from the next lemma:

Lemma 13.2. *Suppose that* $\mathbf{h} : I \mapsto \mathbf{R}^M$ *is integrable (i.e., that* $\int_I \|\mathbf{h}\| \, dt < \infty$*), where* $I \subset \mathbf{R}$ *is an interval. Assume that* $\int_I \mathbf{h} \, dt = 0$*. For all* x *such that* $d(x, I) > \ell(I)$*,*

$$G_\beta(\mathbf{h})(x) \le C(\beta, d)\|\mathbf{h}\|_1 |I|^{-1}(1 + |x - x_I|/\ell(I))^{-1-\beta}, \qquad (13.8)$$

where we are using x_I *to denote* I*'s center.*

We essentially know this. If h is scalar-valued, inequality 6.10 says that

$$G_\beta(h)(x) \le C(\beta, d)\|h\|_1 |I|^{-1}(1 + |x - x_I|/\ell(I))^{-1-\beta}$$

when $d(x, I) > \ell(I)$. Therefore, if $\mathbf{h} = (h_j)_1^M$ is a vector, $\mathbf{G}_\beta(\mathbf{h})$ will be a vector $(P_j)_1^M$, each of whose components P_j is bounded by a constant times

$$\left(\int_I |h_j(t)| \, dt\right) |I|^{-1}(1 + |x - x_I|/\ell(I))^{-1-\beta}.$$

The size of such a vector is dominated by

$$\left(|I|^{-1}(1 + |x - x_I|/\ell(I))^{-1-\beta}\right) \times \left(\sum_j \left(\int_I |h_j(t)| \, dt\right)^2\right)^{1/2}.$$

However (see exercise 13.1),

$$\left(\sum_j \left(\int_I |h_j(t)| \, dt\right)^2\right)^{1/2} \le \int_I \left(\sum_j |h_j(t)|^2\right)^{1/2} dt = \int_I \|\mathbf{h}(t)\| \, dt,$$

which is the conclusion of the lemma. Theorem 13.4 is proved.

Exercises

13.1. Prove inequality 13.1. (Hint: It's enough to show that, if $\mathbf{v}$ is any vector in ℓ^2 with only finitely many non-zero components, then

$$\left| \left(\int_E \mathbf{f} \, dt \right) \cdot \mathbf{v} \right| \leq \|\mathbf{v}\| \int_E \|\mathbf{f}\| \, dt.)$$

13.2. State and prove d-dimensional generalizations of Theorems 13.1–13.4 and Corollaries 13.1 and 13.2.

13.3. The statement of Theorem 13.1 requires that the component functions f_j all be finite sums of Haar functions, and Theorem 13.3 requires that they belong to $\mathcal{C}_0^\infty(\mathbf{R})$. To what extent can these hypotheses be weakened?

14

Random Pointwise Errors

The reader might recall that in an earlier chapter we raised the question of the effect of random errors in summing Haar function expansions. Let us suppose we have a Haar function series,

$$\sum_I \langle f, h_{(I)} \rangle h_{(I)} \equiv \sum_I \lambda_I(f) h_{(I)}, \qquad (14.1)$$

which we wish to sum it up to recover f. The problem we face is that the series we add up won't be 14.1, but

$$\sum_I \lambda_I(f)(1 + \epsilon_I) h_{(I)}, \qquad (14.2)$$

where the ϵ_I's are (we hope) small, but random errors. Set $\epsilon \equiv \sup_I |\epsilon_I|$. Our square function results prove that, if $1 < p < \infty$, then

$$\left\| \sum_I \epsilon_I \lambda_I(f) h_{(I)} \right\|_p \leq C_p \epsilon \|f\|_p,$$

showing that, under reasonable assumptions, the effects of the errors in 14.2 are manageable, at least in a "norm" sense.

What of pointwise summation? Because we want to avoid a detailed discussion of probability theory, we will only consider this problem in a very simple case. We suppose that we have the "true" series $\sum \lambda_I(f) h_{(I)}$, in which each summand is perturbed by an amount equal to $\pm \epsilon \lambda_I h_{(I)}$, where $\epsilon > 0$ is small and '$\pm$' represents a sequence of "fair" (50-50 probabilities) and independent sign changes. What these probabilities and what independence mean both require some explanation.

In the real world, we often do not measure the $\lambda_I(f)$'s once and for all, but repeatedly, and then take averages, with the hope that our measurement errors will mostly cancel out. The question is: How effective—or how rapid—is this cancelation? That is what a probabilistic analysis tries to address.

Each observation of f and its Haar coefficients is bookmarked, so to speak, by a point ω in a probability space Ω. For us Ω will be $[0, 1)$ endowed with Lebesgue measure. Borel measurable functions $X_1, \ldots , X_n$, mapping from $[0, 1)$ to $\mathbf{R}$, are said to be *independent* if, for all Borel sets $E_1, \ldots , E_n$, subsets of $\mathbf{R}$,

$$|\{\omega \in [0, 1) :\ X_i(\omega) \in E_i\ \forall 1 \leq i \leq n\}| = \prod_1^n |\{\omega \in [0, 1) :\ X_j(\omega) \in E_j\}|.$$

$$(14.3)$$

Probabilists do not call measurable functions defined on a probability space "functions": they call them *random variables*. We shall follow their convention.

By an approximation argument (which we leave as an exercise), equation 14.3 can be generalized to the following: If the functions $f_i : \mathbf{R} \mapsto \mathbf{R}$ are bounded Borel measurable functions and the $X_i(\cdot)$ $(1 \leq i \leq n)$ are independent, then

$$\int_0^1 (\prod_1^n f_i(X_i(\omega)))\, d\omega = \prod_1^n \left(\int_0^1 f_i(X_i(\omega))\, d\omega \right).$$

We will be interested in a particular family of independent random variables $\{r_i(\omega)\}_1^\infty$. We will make our discussion easier to follow by first defining these "variables" for all $\omega \in \mathbf{R}$ and then restricting them to $[0, 1)$.

The first *Rademacher function* $r_1(\omega)$ is the 1-periodic function defined on $[0, 1)$ by

$$r_1(\omega) = \begin{cases} 1 & \text{if } 0 \leq \omega < 1/2; \\ -1 & \text{if } 1/2 \leq \omega < 1. \end{cases}$$

For every positive integer n, the nth Rademacher function $r_n(t)$ is defined by $r_n(t) \equiv r_1(2^{n-1}t)$.

For any Borel $E \subset \mathbf{R}$ and any n, the set $\{\omega \in [0, 1) :\ r_n(\omega) \in E\}$ has measure 0, 1, or 1/2. In fact:

$$|\{\omega \in [0, 1) :\ r_n(\omega) \in E\}| = \begin{cases} 0 & \text{if } E \cap \{-1,\, 1\} = \emptyset; \\ 1 & \text{if } \{-1,\, 1\} \subset E; \\ 1/2 & \text{if } E \cap \{-1,\, 1\} \text{ is a singleton.} \end{cases}$$

It is not too difficult to show that the Rademacher functions constitute a family of independent random variables. We strongly recommend that the student verify this for himself: the exercise will teach him a lot about what is going on here.

The Rademacher functions let us quantify the intuitive notion of random, fair-probability changes in sign. Let $\omega \in [0, 1)$, and consider the sequence of numbers $r_{p_1}(\omega), r_{p_2}(\omega), \ldots , r_{p_n}(\omega)$, where $1 \leq p_1 < p_2 < \cdots < p_n$. This is a sequence of ± 1's. There are 2^n possible such sequences, and each one occurs on a set of ω's having probability exactly equal to 2^{-n}. Consider the quantity:

$$|r_{p_1}(\omega) + r_{p_2}(\omega) + \cdots + r_{p_n}(\omega)|. \tag{14.4}$$

If we choose the right ω (or "wrong", depending on how you look at it), this might be as big as n. However, that only happens on a set of probability 2^{-n+1}. We would like to know: What is the average size of 14.4? Actually, we would like to know something a little more precise. Let $\gamma_1, \ldots, \gamma_n$ be real numbers, and consider the quantity

$$|\gamma_1 r_{p_1}(\omega) + \gamma_2 r_{p_2}(\omega) + \cdots + \gamma_n r_{p_n}(\omega)|. \tag{14.5}$$

In order to understand the effects of random errors on summing Haar expansions, we will need to have a good estimate of the average value of 14.5.

Such an estimate is provided by the following theorem.

Theorem 14.1. *For every $0 < p < \infty$, there are positive constants $c_1(p)$ and $c_2(p)$ such that, for all finite linear sums $\sum_1^n \gamma_i r_i(\omega)$,*

$$c_1(p) \left(\sum_1^n |\gamma_i|^2 \right)^{1/2} \leq \left(\int_1^1 \left| \sum_1^n \gamma_i r_i(\omega) \right|^p d\omega \right)^{1/p} \leq c_2(p) \left(\sum_1^n |\gamma_i|^2 \right)^{1/2}. \tag{14.6}$$

The constants $c_1(p)$ and $c_2(p)$ do not depend on n or on the γ_i's.

Remark. These are known as *Khinchin's Inequalities*.

Proof of Theorem 14.1. Without loss of generality, we assume that $\sum_1^n |\gamma_i|^2 = 1$. Consider the function

$$f(\omega) \equiv \sum_1^n \gamma_i r_i(\omega),$$

where $\omega \in [0, 1)$. Here is an easy exercise: If $h_{(I)}$ is any Haar function such that $I \subset [0, 1)$, then there is exactly one index i such that $h_{(I)}$'s inner product with the Rademacher function r_i is non-zero, and that inner product equals $\sqrt{|I|}$. Therefore, $S(f)$, the dyadic square function of f, satisfies

$$S(f)(\omega) = \left(\sum_1^n |\gamma_i|^2 \right)^{1/2}$$

on all of $[0, 1)$. The proof of Theorem 3.2 now implies that, for all $\lambda > 0$,

$$|\{\omega \in [0, 1) : |f(\omega)| > \lambda\}| \leq 2 \exp(-\lambda^2/2); \tag{14.7}$$

and this immediately leads to

$$\int_0^1 |f(\omega)|^p d\omega \leq C_p$$

for all $0 < p < \infty$, which is the right-hand inequality in 14.6.

Hölder's inequality implies the left-hand inequality in 14.6 when $p \geq 2$. Suppose now that $0 < p < 2$. We can write:

$$\sum_1^n |\gamma_i|^2 = \int_0^1 \left| \sum_1^n \gamma_i r_i(\omega) \right|^2 d\omega$$

$$= \int_0^1 \left| \sum_1^n \gamma_i r_i(\omega) \right|^{p/2} \left| \sum_1^n \gamma_i r_i(\omega) \right|^{2-p/2} d\omega$$

$$\leq \left(\int_0^1 \left| \sum_1^n \gamma_i r_i(\omega) \right|^p d\omega \right)^{1/2} \left(\int_0^1 \left| \sum_1^n \gamma_i r_i(\omega) \right|^{4-p} d\omega \right)^{1/2},$$

where the last line follows from Cauchy-Schwarz. However, by the right-hand inequality in 14.6,

$$\left(\int_0^1 \left| \sum_1^n \gamma_i r_i(\omega) \right|^{4-p} d\omega \right)^{1/2} \leq C_p \left(\sum_1^n |\gamma_i|^2 \right)^{1-p/4}.$$

Therefore, after dividing, we get

$$\left(\sum_1^n |\gamma_i|^2 \right)^{p/4} \leq C_p \left(\int_0^1 \left| \sum_1^n \gamma_i r_i(\omega) \right|^p d\omega \right)^{1/2};$$

which, after raising both sides to the $2/p$ power, yields the result.

What do these tell us about pointwise summation? Let's assume that we have a finite Haar function sum with $\pm \epsilon$ errors,

$$\sum_I \lambda_I (1 + \epsilon r_I(\omega)) h_{(I)}(x),$$

where $r_I(\omega)$ is a Rademacher function we have assigned to the dyadic interval I, via some enumeration of $\mathcal{D}$ (in other words, $r_I(\omega) = r_{n_I}(\omega)$). The size of the error at any point x is

$$\epsilon \left| \sum_I \lambda_I r_I(\omega) h_{(I)}(x) \right|,$$

and the expected value of this error—averaging over all $\omega \in [0, 1)$—equals ϵ times

$$\int_0^1 \left| \sum_I r_I(\omega) h_{(I)}(x) \right| d\omega,$$

which is bounded above and below by constants times

$$\left(\sum_I |\lambda_I|^2 (h_{(I)}(x))^2 \right)^{1/2}.$$

But that is simply the dyadic square function of $\sum_I \lambda_I h_{(I)}$!

What have we gained? Our goal is to estimate $\sum_I \lambda_I h_{(I)}(x)$. Unfortunately, the measured value of this sum depends delicately on cancelation of errors in the Haar coefficients and functions. However, as long as ϵ is not too big ($\epsilon < 1/2$ will do), we can get a pretty good estimate of the square function from the observed Haar coefficients $\lambda_I(1 + \epsilon r_I(\omega))$. The square function then gives us a measure of how much confidence we should put in any estimated value of $\sum_I \lambda_I h_{(I)}$. Namely, if the square function is large, we should not expect the sum to be too accurate. Put another way, if the square function is large, we should expect to have to average $\sum_I \lambda_I(1 + \epsilon r_I(\omega))h_{(I)}(x)$ over many observations before getting a trustworthy estimate of $\sum_I \lambda_I h_{(I)}(x)$. The square function also gives us a picture of the amount and type of "spread" we should see in the sums: a standard deviation roughly equal to $S(f)$, with sub-Gaussian decay.

Exercises

14.1. The Rademacher functions provide a fairly easy way to construct a weight v that is dyadic doubling but not in A_∞^d. (The reader might want to look back at Definition 2.4 and Definition 2.5.) Let $0 < \lambda < 1$ and, for N a positive integer, define

$$v_N(t) \equiv \prod_1^N (1 + \lambda r_k(t))$$

for $t \in \mathbf{R}$. a) Show that, for all N and for all dyadic I,

$$1 \leq \frac{v_N(I_l)}{v_N(I_r)} \leq \frac{1+\lambda}{1-\lambda},$$

where I_l and I_r are the right and left halves of I. Show also that $v_N([0,1)) = 1$ for all N. b) Show that, for all $\eta > 0$,

$$\left| \left\{ t \in [0,1) : \frac{1}{N} \sum_1^N r_k(t) < \eta \right\} \right| \to 1 \tag{14.8}$$

as $N \to \infty$. c) Let $\delta > 0$ be so small that $\rho \equiv (1 - \lambda^2)(1 + \lambda)^\delta < 1$. Consider the sequence $\{r_1(t), r_2(t), \ldots, r_N(t)\}$. It consists of m terms equal to -1 and $N - m$ terms equal to $+1$. For such a t, the value of $v_N(t)$ is

$$\left((1 - \lambda^2)(1 + \lambda)^\delta \right)^m (1 + \lambda)^{N-(2+\delta)m},$$

and this will be $\leq \rho^m$ if $N - (2 + \delta)m \leq 0$. Use this fact and 14.8 to show that $v_N \to 0$ in measure on $[0, 1)$. Since $v_N([0, 1)) = 1$, this implies that the sequence $\{v_N\}$, while being uniformly dyadic doubling, is not uniformly in A_∞^d. By restricting v_1 to $[0, 1)$, v_2 to $[1, 2)$, v_3 to $[2, 3)$, and so on, and setting our weight equal to 1 on $(-\infty, 0)$, we get a weight that is globally dyadic doubling on $\mathbf{R}$ but does not belong to A_∞^d.

14.2. How might the preceding construction be extended to $\mathbf{R}^d$ for $d > 1$?

14.3. How might the construction in exercise 14.1 be modified to yield a weight that is doubling but not in A_∞? (Hint: See the reference to Riesz products below.)

Notes

The original proof of Khinchin's Inequalities, which exploited the independence of the Rademacher functions, appeared in [33]. The exponential-square estimate at the heart of the proof was the motivation for Theorem 3.2, which was first conjectured by E. M. Stein. The exercise is a based on a construction I learned from José Luis Fernandez (now at the Universidad Autónoma de Madrid) when we were post-docs at the University of Wisconsin in Madison. The construction can be thought of as a "dyadic Riesz product." If the reader has never seen Riesz products before, he will find a nice introduction to them in [31].

References

1. D. L. Burkholder. Distribution function inequalities for martingales. *Ann. Probability*, 1:19–42, 1973.
2. D. L. Burkholder and R. F. Gundy. Distribution function inequalities for the area integral. *Studia Math.*, 44:527–544, 1972.
3. D. L. Burkholder, R. F. Gundy, and M. L. Silverstein. A maximal function characterization of the class H^p. *Trans. Amer. Math. Soc.*, 157:137–153, 1971.
4. A. P. Calderón. Intermediate spaces and interpolation, the complex method. *Studia Math.*, 24:113–190, 1964.
5. A. P. Calderón. An atomic decomposition of distributions in parabolic H^p spaces. *Advances in Math.*, 25:216–225, 1977.
6. A. P. Calderón and A. Torchinsky. Parabolic maximal functions associated with a distribution, I. *Advances in Math.*, 16:1–64, 1975.
7. A. P. Calderón and A. Torchinsky. Parabolic maximal functions associated with a distribution, II. *Advances in Math.*, 24:101–171, 1977.
8. A. P. Calderón and A. Zygmund. On the existence of certain singular integrals. *Acta. Math.*, 88:85–139, 1952.
9. S. Y. A. Chang and R. Fefferman. A continuous version of duality of H^1 and BMO on the bidisc. *Ann. of Math.*, 112:179–201, 1980.
10. S. Y. A. Chang, J. M. Wilson, and T. H. Wolff. Some weighted norm inequalities concerning the Schroedinger operators. *Comm. Math. Helv.*, 60:217–246, 1985.
11. S. Chanillo and R. L. Wheeden. L^p-estimates for fractional integrals and Sobolev inequalities with applications to Schroedinger operators. 10:1077–1116, 1985.
12. S. Chanillo and R. L. Wheeden. Some weighted norm inequalities for the area integral. *Indiana U. Math. Jour.*, 36:277–294, 1987.
13. R. Coifman and C. Fefferman. Weighted norm inequalities for maximal functions and singular integrals. *Studia Math.*, 51:269–274, 1974.
14. D. Cruz-Uribe and C. Pérez. Two-weight extrapolation via the maximal operator. *J. Funct. Anal.*, 174:1–17, 2000.
15. I. Daubechies. *Ten Lectures on Wavelets*, volume 61 of *CBMS-NSF Regional Conferences in Applied Mathematics*. Society for Industrial and Applied Mathematics, Philadelphia, 1992.
16. J. Duoandikoetxea. *Fourier analysis*. Number 29 in Graduate Studies in Mathematics. American Mathematical Society, Providence, 2001.

17. C. Fefferman. The uncertainty principle. *Bull. Amer. Math. Soc. (NS)*, 9:129–206, 1983.

18. C. Fefferman and E. M. Stein. Some maximal inequalities. *Amer. Jour. of Math.*, 93:107–115, 1971.

19. C. Fefferman and E. M. Stein. H^p spaces of several variables. *Acta Math.*, 129:137–193, 1972.

20. R. Fefferman, R. F. Gundy, M. L. Silverstein, and E. M. Stein. Inequalities for ratios of functionals of harmonic functions. *Proc. Nat. Acad. Sci. USA*, 79:7958–7960, 1982.

21. G. B. Folland. *Real Analysis: Modern Techniques and Their Applications*. Wiley Interscience, New York, 1999.

22. M. Frazier, B. Jawerth, and G. Weiss. *Littlewood-Paley Theory and the Study of Function Spaces*. Number 79 in CBMS Regional Conference Series in Mathematics. American Mathematical Society, Providence, 1991.

23. J. Garcia-Cuerva. An extrapolation theorem in the theory of A_p weights. *Proc. Amer. Math. Soc.*, 87:422–426, 1983.

24. J. Garcia-Cuerva and J. L. Rubio de Francia. *Weighted Norm Inequalities and Related Topics*. North-Holland, Amsterdam, 1985.

25. J. B. Garnett. *Bounded Analytic Functions*. Academic Press, New York, 1981.

26. R. F. Gundy and R. L. Wheeden. Weighted integral inequalities for the nontangential maximal function, Lusin area integral, and Walsh-Paley series. *Studia Math.*, 49:107–124, 1973/74.

27. A. Haar. Zur Theorie der orthogonalen Funktionensysteme. *Math. Ann.*, 69:331–371, 1910.

28. G. H. Hardy and J. E. Littlewood. A maximal theorem with function-theoretic applications. *Acta Math.*, 54:81–116, 1930.

29. L. Hormander. Estimates for translation invariant operators in L^p spaces. *Acta Math.*, 104:93–139, 1960.

30. R. A. Hunt, B. Muckenhoupt, and R. L. Wheeden. Weighted norm inequalities for the conjugate function and Hilbert transform. *Trans. Amer. Math. Soc.*, 176:227–251, 1973.

31. Y. Katznelson. *An Introduction to Harmonic Analysis*. Dover, New York, 1976.

32. R. Kerman and E. Sawyer. The trace inequality and eigenvalue estimates for Schroedinger operators. *Annales de L'Institut Fourier*, 36:207–228, 1986.

33. A. Khinchin. Ueber dyadische Brueche. *Math. Zeit.*, 18:109–116, 1923.

34. M. A. Krasnosel'skii and Ya. B. Rutickii. *Convex functions and Orlicz spaces*. P. Noordhoff, Groningen, 1961.

35. D. S. Kurtz. Littlewood-Paley and multiplier theorems on weighted L^p spaces. *Trans. Amer. Math. Soc.*, 259:235–254, 1980.

36. D. S. Kurtz and R. L. Wheeden. Results on weighted norm inequalities for multipliers. *Trans. Amer. Math. Soc.*, 255:343–362, 1979.

37. J. E. Littlewood and R. E. A. C. Paley. Theorems on Fourier series and power series, Part I. *J. London Math. Soc.*, 6:230–233, 1931.

38. J. E. Littlewood and R. E. A. C. Paley. Theorems on Fourier series and power series, Part II. *Proc. London Math. Soc.*, 42:52–89, 1937.

39. J. E. Littlewood and R. E. A. C. Paley. Theorems on Fourier series and power series, Part III. *Proc. London Math. Soc.*, 43:105–126, 1937.

40. J. Marcinkiewicz. Sur l'interpolation d'opérations. *C. R. Acad. Sci. Paris*, 208:1272–1273, 1939.

41. S. G. Mihlin. On the multipliers of Fourier integrals. *Dokl. Akad. Nauk.*, 109:701–703, 1956.

42. B. Muckenhoupt. Weighted norm inequalities for the Hardy maximal function. *Trans. Amer. Math. Soc.*, 165:207–226, 1972.

43. T. Murai and A. Uchiyama. Good-λ inequalities for the area integral and the nontangential maximal function. *Studia Math.*, 83:251–262, 1986.

44. F. Nazarov. Private communication.

45. R. O'Neill. Fractional integration in Orlicz spaces. *Trans. Amer. Math. Soc.*, 115:300–328, 1963.

46. C. Pérez. Weighted norm inequalities for singular integral operators. *J. London Math. Soc.*, 49:296–308, 1994.

47. C. Pérez. On sufficient conditions for the boundedness of the Hardy-Littlewood maximal operator between weighted L^p spaces with different weights. *Proc. London Math. Soc.*, 71:135–157, 1995.

48. C. Pérez. Sharp weighted L^p weighted Sobolev inequalities. *Annales de L'Institut Fourier*, 45:809–824, 1995.

49. J. L. Rubio de Francia. Factorization theory and A_p weights. *Amer. Jour. of Math.*, 106:533–547, 1984.

50. C. Segovia and R. L. Wheeden. On weighted norm inequalities for the Lusin area integral. *Trans. Amer. Math. Soc.*, 176:103–123, 1973.

51. E. M. Stein. On the functions of Littlewood-Paley, Lusin, and Marcinkiewicz. *Trans. Amer. Math. Soc.*, 88:430–466, 1958.

52. E. M. Stein. On some functions of Littlewood-Paley and Zygmund. *Bull. Amer. Math. Soc.*, 67:99–101, 1961.

53. E. M. Stein. *Singular Integrals and Differentiability Properties of Functions.* Princeton University Press, Princeton, 1970.

54. E. M. Stein. The development of square functions in the work of A. Zygmund. *Bull. Amer. Math. Soc.*, 7:359–376, 1982.

55. J. O. Stromberg and A. Torchinsky. *Weighted Hardy Spaces*, volume 1381 of *Lecture Notes in Mathematics.* Springer-Verlag, Berlin, 1989.

56. A. Torchinsky. *Real-Variable Methods in Harmonic Analysis.* Academic Press, New York, 1986.

57. A. Uchiyama. A constructive proof of the Fefferman-Stein decomposition of $BMO(R^n)$. *Acta Math.*, 148:215–241, 1982.

58. G. Weiss. A note on Orlicz spaces. *Portugal Math.*, 15:35–47, 1950.

59. J. M. Wilson. The intrinsic square function. To appear in *Revista Matematica Iberoamericana.*

60. J. M. Wilson. A sharp inequality for the square function. *Duke Math. Jour.*, 55:879–888, 1987.

61. J. M. Wilson. Weighted inequalities for the dyadic square function without dyadic A_∞. *Duke Math. Jour.*, 55:19–49, 1987.

62. J. M. Wilson. Weighted norm inequalities for the continuous square function. *Trans. Amer. Math. Soc.*, 314:661–692, 1989.

63. J. M. Wilson. Chanillo-Wheeden inequalities for $0 < p \leq 1$. *J. London Math. Soc.*, 41:283–294, 1990.

64. J. M. Wilson. Some two-parameter square function inequalities. *Indiana U. Math. Jour.*, 40:419–442, 1991.

65. J. M. Wilson. Paraproducts and the exponential square class. *Jour. of Math. Analy. and Applic.*, 271:374–382, 2002.

66. A. Zygmund. On certain integrals. *Trans. Amer. Math. Soc.*, 55:170–204, 1944.

Index

4. Careful preparation of the manuscripts will help keep production time short besides ensuring satisfactory appearance of the finished book in print and online. After acceptance of the manuscript authors will be asked to prepare the final LaTeX source files (and also the corresponding dvi-, pdf- or zipped ps-file) together with the final printout made from these files. The LaTeX source files are essential for producing the full-text online version of the book (see http://www.springerlink.com/openurl.asp?genre=journal&issn=0075-8434 for the existing online volumes of LNM).

   The actual production of a Lecture Notes volume takes approximately 8 weeks.

5. Authors receive a total of 50 free copies of their volume, but no royalties. They are entitled to a discount of 33.3 % on the price of Springer books purchased for their personal use, if ordering directly from Springer.

6. Commitment to publish is made by letter of intent rather than by signing a formal contract. Springer-Verlag secures the copyright for each volume. Authors are free to reuse material contained in their LNM volumes in later publications: A brief written (or e-mail) request for formal permission is sufficient.

Addresses:

Professor J.-M. Morel, CMLA,
École Normale Supérieure de Cachan,
61 Avenue du Président Wilson, 94235 Cachan Cedex, France
E-mail: Jean-Michel.Morel@cmla.ens-cachan.fr

Professor F. Takens, Mathematisch Instituut,
Rijksuniversiteit Groningen, Postbus 800,
9700 AV Groningen, The Netherlands
E-mail: F.Takens@math.rug.nl

Professor B. Teissier, Institut Mathématique de Jussieu,
UMR 7586 du CNRS, Équipe "Géométrie et Dynamique",
175 rue du Chevaleret
75013 Paris, France
E-mail: teissier@math.jussieu.fr

For the "Mathematical Biosciences Subseries" of LNM:

Professor P. K. Maini, Center for Mathematical Biology,
Mathematical Institute, 24-29 St Giles,
Oxford OX1 3LP, UK
E-mail : maini@maths.ox.ac.uk

Springer, Mathematics Editorial, Tiergartenstr. 17,
69121 Heidelberg, Germany,
Tel.: +49 (6221) 487-8410
Fax: +49 (6221) 487-8355
E-mail: lnm@springer-sbm.com

Printed by Books on Demand, Germany